Methods in Neuroethological Research

Hiroto Ogawa • Kotaro Oka
Editors

Methods in Neuroethological Research

Editors

Hiroto Ogawa, Dr.
Department of Biological Science
Faculty of Science
Hokkaido University
Kita 10, Nishi 8, Kita-ku
Sapporo 060-0810, Japan
hogawa@sci.hokudai.ac.jp

Kotaro Oka, Dr.
Department of Biosciences and Informatics
Faculty of Science and Technology
Keio University
3-14-1 Hiyoshi, Kohoku-ku
Yokohama 223-8522, Japan
oka@bio.keio.ac.jp

ISBN 978-4-431-54330-5   ISBN 978-4-431-54331-2 (eBook)
DOI 10.1007/978-4-431-54331-2
Springer Tokyo Heidelberg New York Dordrecht London

Library of Congress Control Number: 2013943814

© Springer Japan 2013

This work is subject to copyright. All rights are reserved by the Publisher, whether the whole or part of the material is concerned, specifically the rights of translation, reprinting, reuse of illustrations, recitation, broadcasting, reproduction on microfilms or in any other physical way, and transmission or information storage and retrieval, electronic adaptation, computer software, or by similar or dissimilar methodology now known or hereafter developed. Exempted from this legal reservation are brief excerpts in connection with reviews or scholarly analysis or material supplied specifically for the purpose of being entered and executed on a computer system, for exclusive use by the purchaser of the work. Duplication of this publication or parts thereof is permitted only under the provisions of the Copyright Law of the Publisher's location, in its current version, and permission for use must always be obtained from Springer. Permissions for use may be obtained through RightsLink at the Copyright Clearance Center. Violations are liable to prosecution under the respective Copyright Law.

The use of general descriptive names, registered names, trademarks, service marks, etc. in this publication does not imply, even in the absence of a specific statement, that such names are exempt from the relevant protective laws and regulations and therefore free for general use.

While the advice and information in this book are believed to be true and accurate at the date of publication, neither the authors nor the editors nor the publisher can accept any legal responsibility for any errors or omissions that may be made. The publisher makes no warranty, express or implied, with respect to the material contained herein.

Printed on acid-free paper

Springer is part of Springer Science+Business Media (www.springer.com)

Preface

Neuroscience is one of the most multidisciplinary research fields, employing a great variety of research approaches. To conduct investigation in neuroscience, researchers need to understand a wide variety of experimental and analytical methods and to make full use of a broad range of techniques. In addition, they have to use peculiar techniques such as electrophysiology that are unfamiliar in other fields of life science studies. In some instances, at least modest computational skill for mathematical analysis and modeling is required. Researchers, including ourselves, tend to persist in familiar methods and to have reservations about the introduction of novel approaches. However, manifold methods are strongly required for attaining a higher level of research products in this field. Further, because technological progress is much faster now, researchers must work to obtain information not only on the specific field of their own study but also on methodology, and must continue to renew their research techniques to keep them up-to-date.

Actually, it is impossible for a single researcher to be familiar with all of the methods. In many cases, collaborative work between specialists with different techniques reduces the strain on each other. Even in this case, it goes without saying that we should have a fundamental knowledge of the techniques used for our own research work. Now, we can get a large number of cookbooks on various techniques, recipes, and protocols used for neuroscience. However, most of these books provide a special technique or detail of individual methods in specific model animals, while there are very few books reviewing overall methodology used for neuroscience research. For that reason, we came up with a book plan that provided a good starting point for neuroscience students, newcomers, and young researchers to consider the introduction of new experimental strategies, especially for neuroethology.

The genesis of this book plan was in our organization of a symposium entitled "Strategy for the New Generation of Neuroethology" at the annual meeting of the Japan Neuroscience Society in 2007. Neuroethology is a branch of neuroscience that seeks to understand the neural basis of *natural* animal behavior. Neuroethological studies have covered various types of behaviors in a wide diversity of animals including locusts, crickets, honeybees, electric fish, toads, sea turtles, barn owls,

Fig. 1 Behavioral control with optogenetic tools. See also Chap. 8

bats, and others. In addition to these classic subjects, so-called model animals such as *Caenorhabditis elegans*, *Drosophila,* and zebrafish are generally used for specimens in neuroethological studies because they provide a major advantage for molecular genetic techniques, of which application and variation has expanded rapidly. In 2007, neuroethological research using model animals was not yet very popular. However, we predicted that optical imaging with genetic probes and genetic manipulation of neural activity would become important tools in the next 10 years if they were combined with conventional methods such as electrical physiology and behavior observation. We therefore dared to select topics investigating model animals in the organization of the symposium. This symposium with its new initiative acquired a good reputation, and Springer Japan suggested to us the creation of a book plan related to the symposium. Every reader knows that a number of studies focused on model animals have had great impacts on the neuroscience field in the past several years.

The aim of this book is to introduce novel experimental and analytical techniques useful for present and upcoming neuroethological study to students and young researchers. Therefore, we did not list traditional approaches but, rather, methods unfamiliar in neuroethology with selected applications for *C. elegans* (Chap. 1), *Drosophila* (Chap. 7), and mice (Chap. 8). In particular, optical imaging using the genetic probe and optogenetics that have still been adopted only in the model animals would also become a powerful tool for other animals used in neuroethology (Fig. 1). On the other hand, we also took up applications of the new technology to conventional neuroethological materials such as honeybees (Chaps. 2 and 10), crickets (Chap. 5), and earthworms (Chap. 6). These cases show the possibility that advanced approaches could provide new findings even in unpopular, non-model animals that have disadvantages in molecular genetics.

Furthermore, these chapters will serve as useful references encouraging ingenuity or providing know-how in applying the new techniques to specific animals with unique behavior. Electrophysiology, which directly records neural activities, although conventional, remains an important method in neuroethology. Recent advanced analytical methods derived from electrophysiological data can clarify complex spatio-temporal pattern encoding environmental information to build a simulation model of neural processing (Chaps. 3 and 4). Especially, the MATLAB toolbox is very powerful for on- and offline analysis of electrophysiological data recorded in various species of animals. Finally, spatial analysis of the expression pattern of the immediate early gene is widely used for the monitoring of neural activity. For example, this method has provided important results on the neural system underlying song learning in songbirds (Chap. 9). Recent molecular-biological approaches demonstrate that the gene expression induced by learning and experience is epigenetically regulated in vertebrates and invertebrates (Chap. 10).

Most of the chapters in this book focus not on original innovation of novel techniques but on how to apply those methods to a particular research theme or experimental animal. In fact, it is more common for researchers not to develop their own original methods independently but to use ingenuity and innovation in employing established methods for their studies. This application is not such an easy process for researchers who investigate the unique natural behavior of non-model animals in neuroethology. A great effort is required for modification of methodology to adjust it to different animals and research themes if reliable and significant results are to be obtained. The young researchers we asked to write each chapter of this book have tried to create novel methods of their own and to obtain noteworthy progress in their research. We hope that this book will encourage many students and postdocs in neuroethological research to try new approaches. We will be greatly pleased if this book can support their research work.

Hiroto Ogawa
Kotaro Oka

Contents

Part I Behavioral Analysis

1. **Behavioral Analysis in *Caenorhabditis elegans*** 3
   Yuki Tsukada and Ikue Mori

2. **Classical Conditioning of the Proboscis Extension Reflex in the Honeybee** 15
   Yukihisa Matsumoto, Jean-Christophe Sandoz, and Martin Giurfa

Part II Electrophysiology

3. **Mining Spatio-Spectro-Temporal Cortical Dynamics: A Guideline for Offline and Online Electrocorticographic Analyses** 39
   Zenas C. Chao and Naotaka Fujii

4. **Computational Analysis of Behavioural and Neural Data Through Bayesian Statistical Modelling** 57
   Raymond Wai Mun Chan and Fabrizio Gabbiani

Part III Optical Recording Techniques

5. **In Vivo Ca^{2+} Imaging of Neuronal Activity** 71
   Hiroto Ogawa and John P. Miller

6. **Optical Imaging Techniques for Investigating the Function of Earthworm Nervous System** 89
   Kotaro Oka and Hiroto Ogawa

Part IV Optogenetics

7 Monitoring Neural Activity with Genetically Encoded Ca^{2+} Indicators 103
Azusa Kamikouchi and André Fiala

8 Controlling Behavior Using Light to Excite and Silence Neuronal Activity 115
Ali Cetin and Shoji Komai

Part V Molecular Biological Techniques

9 Detecting Neural Activity-Dependent Immediate Early Gene Expression in the Brain 133
Kazuhiro Wada, Chun-Chun Chen, and Erich D. Jarvis

10 Epigenetic Regulation of Gene Expression in the Nervous System 151
Dai Hatakeyama, Sascha Tierling, Takashi Kuzuhara, and Uli Müller

Index 173

Part I
Behavioral Analysis

Chapter 1
Behavioral Analysis in *Caenorhabditis elegans*

Yuki Tsukada and Ikue Mori

Abstract Behavior is an eventual output of animals' responses to environmental stimuli and is also an outcome of highly orchestrated mechanisms with different levels. Understanding the principles of behavioral regulation thus requires the illumination of such mechanisms from multiple angles. Model animals provide practical solutions for this requirement by accumulating different aspects of knowledge about focused animal. In this chapter, we introduce behavioral analyses for *Caenorhabditis elegans* (*C. elegans*), the simplest multicellular model animal, which is useful for various kinds of studies. We can incorporate different kinds of approaches to shed light on the mechanisms of behavioral regulation by using this convenient model animal. This chapter presents description of typical population analysis, single animal analysis, use of genetically encoded calcium indicators (GECI), and combination of these approaches for the study of *C. elegans* behavior. Computerized automated methods are also mentioned for efficient experimental design.

Keywords Behavioral analysis • *C. elegans* • *Caenorhabditis elegans* • Calcium imaging • Computerized methods • Genetically encoded calcium indicators • Population analysis

Y. Tsukada (✉) • I. Mori (✉)
Division of Biological Science, Graduate School of Science,
Nagoya University, Furou-cho, Chikusa-ku, Nagoya 464-8602, Japan
e-mail: tsukada.yuki@nucc.cc.nagoya-u.ac.jp; m46920a@nucc.cc.nagoya-u.ac.jp

1.1 Introduction: *C. elegans* as a Model Organism for Behavioral Analysis

About 1 mm long, tiny nematode *Caenorhabditis elegans* has been used for biological research as an important model organism for nearly 40 years (Brenner 1974). An abundant advantage for scientific studies such as short life cycle, transparent body, and convenience of genetic analysis has been pushing various fields of research including genomics, cell biology, development, aging, and neuroscience.

Since behavior is an eventual output of biological system, accumulated detailed knowledge about *C. elegans* tremendously contributes to uncover the mechanisms of behavioral regulation. Complete anatomical dissection exhibited connectivity and identity of all 302 neurons for an adult hermaphrodite and 383 neurons for adult male; furthermore, we can easily search such anatomical map by using online database (Hunt-newbury et al. 2007). Together with genetic database that provides genome information (Harris et al. 2010), abundant information at cellular, molecular, and genetic levels related to acquired behavioral analysis can be obtained. Hence, behavioral analysis for *C. elegans* enables to connect molecular, cellular, or genetic levels of mechanism for behavioral regulation.

Additional important characteristic to use *C. elegans* for current biological research is an excellent affinity for optical methods including imaging and optogenetics. Transparent body already has been afforded powerful methodology for dissecting cellular properties: combination with differential interference contrast (DIC) microscope or genetically labeled fluorescent probes such as green fluorescent protein (GFP) is a prominent case to show experimental merits of *C. elegans*. In addition, GECIs such as Cameleons (Kimura et al. 2004; Miyawaki et al. 1997; Nagai et al. 2004) or GCaMPs (Nakai et al. 2001; Tian et al. 2009) are key tools for exploring relationship between activity of specific neurons and particular behavior. Transparent body is highly compatible for such imaging probes, and thus making these probes enable to observe neural activity and behavior simultaneously. Recently, several kinds of opsin genes are utilized to probe neuronal functions by specific perturbation of cellular ionic current with light (Yizhar et al. 2011). In such optogenetic approach of neural networks, *C. elegans* keeps its superiority with its plenty of cell-specific promoters, a complete connection map for all neurons.

As mentioned shortly above, there are immense advantages to use *C. elegans* for behavioral studies. In this chapter, we describe typical behavioral analysis for *C. elegans* that connects results of genetics, imaging, and other physiological data. The following sections consist of (1) population analysis, (2) single animal analysis, (3) calcium imaging with GECIs, and (4) simultaneous monitoring for behavior and neural activity with highly advanced techniques.

1.2 Population Analysis

Behavioral data often contains stochastic or probabilistic components, which frequently mask essential characters in observed phenomena. We should then design behavioral experiments to detect the essence of observed phenomena in spite of

1 Behavioral Analysis in *C. elegans*

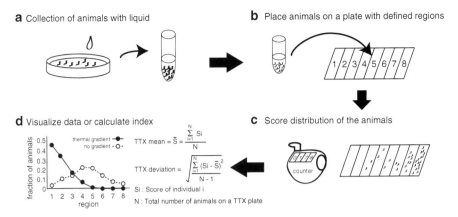

Fig. 1.1 Protocol of population behavioral analysis for *C. elegans*. (**a**) Cultivated animals were collected and washed with liquid. Several buffers were used depending on the purpose of experiments. (**b**) Collected animals were placed on assay plate. Extra liquid were removed by tissue paper to prevent the worms from swimming. (**c**) Score the distribution of animals in the assay plate after certain assay time. (**d**) Visualize scored data and/or calculate characteristic index. Statistical analysis should be needed to discriminate the difference between the different conditions

probabilistic property of animal behavior. In this section, we describe about population assay for *C. elegans* in which we use relatively a large number of animals for each experiment. Such assay systems enable to provide well-reproducible data even with rather small number of trials. Due to small body size and prolific nature, population assay for *C. elegans* is easy and practical in laboratory experiments. Thus, the population analysis for *C. elegans* can reduce time and cost to validate the focusing behavioral character, as compared with single animal analysis described in the later section.

1.2.1 Protocol of Population Analysis

Exact protocols of population assay vary depending on its purpose. We thus summarize here four main steps of population assay shown in Fig. 1.1. First, a large number of animals are collected from cultivating plate (Fig. 1.1a). Liquids such as wash buffer (0.02 % gelatin, 50 mM NaCl, and 25 mM pH 6.0 potassium phosphate) or NG buffer (0.3 % NaCl, 1 mM $CaCl_2$, 1 mM $MgSO_4$, and 25 mM pH 6.0 potassium phosphate) are used for this collection step. Note that the buffer composition sometimes affects behavioral assay so that the selection of buffer should be carefully considered. Usually this collection step includes aiming to wash off the food sticking to the animal body. Second, the collected animals are gently placed on an assay plate with defined regions (Fig. 1.1b). The assay plate should be kept with constant and stable environment during assay. Since behavior is affected by any kinds of stimuli around the plate, experimenters should be attentive to keep the assay environment strictly controlled in fine range. Third, animals in each region are scored after a certain time (Fig. 1.1c). Scoring should be conducted with strict categories of animals based on developmental stage, marker gene expression, or any detectable characters, for example, scoring only adult animals. Then, behavioral characteristics could be recognized with certain data visualization (Fig. 1.1d). Calculating index is helpful to discriminate different sets of animals, and several statistical methods are used to certify the difference between each set. We should especially be careful about statistical test for population assay when we perform nonconventional way, because sometimes the appropriate statistical methods for population assay are complicated, and inappropriate statistical methods make the results distorted. Clearly, controls are necessary for behavioral assay because behavioral responses often vary from day to day. Both negative and positive controls should be contained in experimental designs.

1.2.2 Analysis for Thermotaxis as an Example of Population Assay

One of the prominent examples of population analysis for *C. elegans* is thermotaxis assay. Thermotaxis is observed when we grow *C. elegans* at a constant temperature around 16–25 °C with plenty of food and put them on the plate with thermal gradient without food. The behavioral responses to the environmental temperature were originally found and defined as thermotaxis in 1975 (Hedgecock and Russell 1975); it has been considered as one of the good model systems to elucidate behavioral regulation by neural networks because of its simple neural network to drive the thermotaxis (Mori and Ohshima 1995) and powerful genetic methods of *C. elegans* (Mori 1999). Population analysis is used for examining genetic, conditional, and pharmacological effects (Ito et al. 2006; Jurado et al. 2010; Nishida et al. 2011; Ohnishi et al. 2011; Sugi et al. 2011). In a typical thermotaxis assay, animals conditioned with constant temperature are placed on a thermal gradient, and the animals show various behaviors depending on conditioning, mutation, thermal environment, and other factors (Fig. 1.1d).

1.3 Single Animal Analysis

Population analysis is not fit to observe particular types of animal behavior. For instance, a study of touch response for animals requires observation of immediate action against touch stimulus. To quantify such behavioral responses, single animal observation works well. It is usually performed with stereomicroscope after the isolation of single animal, but a trail on agar plate could be also used to identify typical behavior without microscopes (Fig. 1.2a).

1 Behavioral Analysis in *C. elegans*

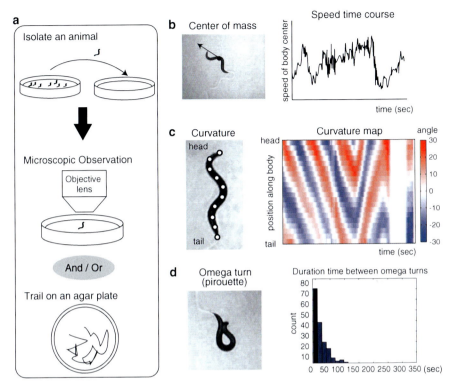

Fig. 1.2 Examples of single worm analysis. (**a**) Typical protocol of single animal assay: first isolate a cultivated animal to an agar plate, then observe with a microscope and/or picture a trail of the animal on agar plate after certain time. (**b**) Migrating speed is usually measured with center of body mass. (**c**) Curvature map along body axis clearly shows kinematic property of *C. elegans* behavior. White square region in the map denotes curled or turning bend of the animal detected by another recognition algorithm. (**d**) Distribution of characteristic behavior such as turning is often used for depicting behavior

1.3.1 Conventional Methods of Single Animal Observation

Depicting with a stereomicroscope is the most sensitive and conventional way for behavioral observation of *C. elegans*. A researcher can recognize different styles of animal response to given stimulus and can compare quantity of particular action among mutants, for instance, tapping the plate provokes backward motion of *C. elegans*; comparison of frequency and distance of the backward motion in response to tapping stimulus was classically performed by human observation to measure habituation. The critical step of such a human judge-based analysis is definition of each behavior for scoring. For example, in the case of reversal behavior, there are several types of reversals including short quick reversals or long reversal with directional change. Therefore, precise definition of identifying behavior should be used consistently among experimenters and publications.

Another way of observation for individual *C. elegans* behavior is the use of a trail on an agar plate. With a certain condition of agar plate, a trail of freely moving animal could be identified and recorded in a photo picture. Isothermal tracking in thermotaxis is a prominent example for using this method (Biron et al. 2006; Mori and Ohshima 1995).

1.3.2 Computerized Methods

Computer-based automated regulation and image-processing methods promote efficiency of assays for *C. elegans* and digging out the essence of behavioral components. Combination of motorized stage with image-processing systems or tracking a freely moving animal has become already conventional for recording movies of *C. elegans* behavior (Hoshi and Shingai 2006; Ramot et al. 2008a). Exploiting such computational systems enabled to obtain various kinds of characters that never have been otherwise obtained in a quantitative manner. Migration speed is one of dynamic properties that reflect the internal state of an animal. Speed is usually defined using the center of target animal's body, namely, center of mass of the animal in a binary image (Fig. 1.2b). Since speed data often contain fluctuation caused by its definition, one should consider managing the fluctuation. Similar to the approach in Sect. 1.5, combination of speed and neuronal activity uncovers functions of specific neuron or molecule (Kawano et al. 2011). *C. elegans* uses undulatory locomotion for migration. When we measure curvature along anterior-posterior axis, patterns of curvature change exhibit particular behaviors: flows of a bend from head to tail denote forward movement, and conversely, reverse flows denote backward movement (Fig. 1.2c). This curvature map can capture speed of locomotion, and it also enables to identify typical patterns of movement like swimming (Fang-Yen et al. 2010; Pierce-Shimomura et al. 2008). Vast numbers of quantitative measurement can be acquired by automated methods; it enables to illuminate the mechanisms from macro view. For example, histogram of duration time between turns shows exponential decrease (Fig. 1.2d, Miyara et al. 2011). These data would be useful for establishing mathematical models because reliably fitted data support hypothesis to construct a mathematical model. Thus, automated acquisition for quantitative data has possibility to accelerate modeling studies and it may lead to draw the integrated view of whole behavioral regulation mechanisms.

1.4 Use of GECIs

GECIs are highly compatible with *C. elegans* studies because of transparent body, exhaustive anatomical information, and abundant cell-specific promoters. Indeed, the first use of fluorescent protein for multicellular organism has established with *C. elegans* (Chalfie et al. 1992). Two calcium probes are mainly used for *C. elegans* studies because of its high sensitivity. These two, Cameleons and GCaMPs, have

1 Behavioral Analysis in *C. elegans*

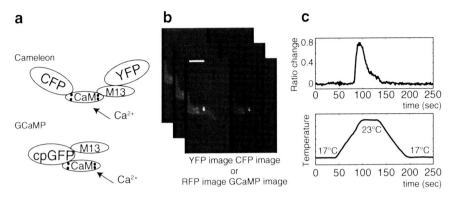

Fig. 1.3 (a) Two of major GECIs: Cameleon and GCaMP. Both contain fluorescent protein(s), calcium-binding sites (*black dots*) of calmodulin, and calmodulin-binding site M13. (b) Simultaneous dual fluorescent imaging is usually required for calcium imaging. Cell-specific expression of the indicators enables to detect single neuronal signal. Thermosensory neuron AFD is pictured here. Scale bar indicates 100 μm. (c) An example of temperature-responding activity of AFD thermosensory neuron. In this case, YFP/CFP increase with temperature increase

several improved variants and further the improvement of such calcium indicators is still actively promoted because of its utility. We shortly describe about these two GECIs for *C. elegans* studies.

1.4.1 Principles of GECIs

Different principles are implemented in Cameleons and GCaMPs. Cameleons consist of two fluorescent proteins, calcium-binding domains of calmodulin, and calmodulin-binding peptide called M13 (Fig. 1.3a, Miyawaki et al. 1997). Four Ca^{2+} binding sites of calmodulin cause conformational change by interacting with M13, and thus, relative orientation between the two fluorescent proteins such as cyan fluorescent protein (CFP) and yellow fluorescent protein (YFP) changes. This relative orientation of the fluorescent proteins affects to energy transfer from CFP to YFP called Förster Resonance Energy Transfer (FRET). Eventually, monitoring intensity of YFP/CFP (Fig. 1.3b; sometimes differently calculated) or fluorescence lifetime reflects concentration of Ca^{2+} in the site where a probe expressed. In the case of GCaMPs, circularly permutated GFP changes intensity of fluorescence emission depending on the Ca^{2+} bindings of the calmodulin domain (Fig. 1.3a, Nakai et al. 2001). Improved GCaMP called GCaMP3 provides high sensitivity and fast response for Ca^{2+} concentration (Tian et al. 2009); it is becoming quickly popular in the studies of *C. elegans*. To compensate the artifact caused by the animal movement, microscopic focal plane change, and photobleaching, another fluorescent protein with different emission wavelength such as red fluorescent protein (RFP) is often used (Fig. 1.3b).

1.4.2 An Example of Calcium Imaging with GECI

We can selectively monitor the neuronal Ca^{2+} change by cell specifically expressing GECIs using cell-specific promoters. In the many cases of *C. elegans* neurons, Ca^{2+} change is thought to be reliably reflecting neural activity because *C. elegans* has no voltage-gated sodium channel. Figure 1.3c shows response of thermosensory neuron AFD to thermal stimulus. In this case, 23 °C cultivated animal responds at just before 23 °C with increase of temperature. YFP/CFP value is calculated with defined neuronal region in the fluorescence images, and average intensity among the region is used for YFP/CFP. Slight differences of the value calculation protocol cause slight change of the results but seldom produce qualitatively different results. However, imaging with GECIs sometimes contains artifact caused by several reasons such as photobleach, the animal movement, and microscopic focal plane change; therefore, experimenters should be careful for these possible artifacts when acquiring image data. Controls always should be accompanied with imaging experiments; in the case of FRET-based GECIs, accepter bleaching or inspection of ratiometric intensity change between two fluorescence proteins is useful to validate FRET occurrences.

1.5 Simultaneous Monitoring for Behavior and Neural Activity

Combination of individual tracking and calcium imaging enables to monitor simultaneous monitoring for stimulus, neural activity, and behavior (Ben Arous et al. 2010; Clark et al. 2007; Piggott et al. 2011). To conduct such simultaneous monitoring, fully or almost computerized system is necessary. Figure 1.4a shows an example of the simultaneous monitoring for thermotaxis. The system consists of thermal gradient, tracking, and Ca^{2+} imaging component. Each component is coordinately controlled with computers. The system records trail of freely moving animal on thermal gradient with time stamps (Fig. 1.4b), thereby enabling the calculation of the time course of temperature stimulus for freely moving animal (Fig. 1.4c). From time-lapse fluorescence images of GECIs, time course of Ca^{2+} concentration change can be obtained as time course of YFP/CFP (Fig. 1.4d). Note that Ca^{2+} imaging for moving object tends to include several kinds of artifacts, which essentially originated from the movement. We should thus particularly be careful for the artifact in the case of Ca^{2+} imaging for moving objects. Integration of Ca^{2+} imaging data and migrating trail comprehensively depict the relationship between the monitoring neural activity and the corresponding behavior. Figure 1.4e shows the correlation between high activity of AFD thermosensory neuron and turning, or straight migration and low activity of AFD. Acquirement and detailed analysis of these quantitative data are powerful for dissecting dynamic phenomena. Moreover, these data can be reconciled with mathematical modeling and thus profitable to understand the mechanisms for regulation of dynamic systems such as animal behavior.

1 Behavioral Analysis in *C. elegans*

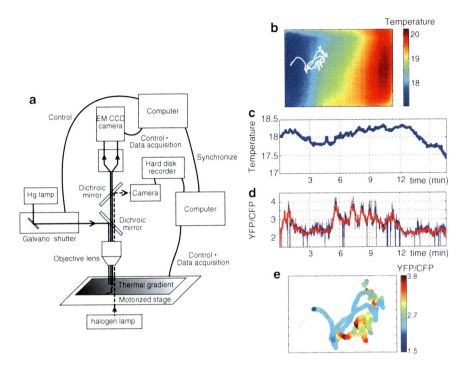

Fig. 1.4 An example of simultaneous measurements for behavior, thermal stimulus, and neuronal activity. (**a**) Tracking and imaging system which combine calcium imaging and automated stage regulation. (**b**) A trail of *C. elegans* (*white line*) on an agar plate with thermal gradient. (**c**) Temperature time course calculated by the data shown in (**b**). (**d**) Time course of YFP/CFP of Cameleon expressed in AFD thermosensory neuron. (**e**) Mapping of YFP/CFP in (**d**) on the trail of animal shown in (**b**)

1.6 Conclusion

Many advantages of *C. elegans* studies are utilized for behavioral analysis as introduced in this chapter. In addition, several other approaches are available and have been developed enthusiastically. Same as *Drosophila*, zebra fish, and mouse research, technique of electrophysiology could be also used for *C. elegans* (Piggott et al. 2011; Ramot et al. 2008b). Recent progress of microfluidic devices enables to expand experimental designs; it could be used for both population (Albrecht and Bargmann 2011) and individual (Chronis et al. 2007) assay according to the research context and particularly useful for the experiments related to odorant sensing. Optogenetics is most suitable with *C. elegans* studies because of transparent body and compatibility with genetics (Ezcurra et al. 2011; Kuhara et al. 2011; Lindsay et al. 2011; Narayan et al. 2011). Using these updating techniques efficiently, studies of *C. elegans* continue to promote neuroethology.

References

Albrecht DR, Bargmann CI (2011) High-content behavioral analysis of *Caenorhabditis elegans* in precise spatiotemporal chemical environments. Nat Methods 8:599–605. doi:10.1038/nmeth.1630

Ben Arous J, Tanizawa Y, Rabinowitch I et al (2010) Automated imaging of neuronal activity in freely behaving *Caenorhabditis elegans*. J Neurosci Methods 187:229–234. doi:10.1016/j.jneumeth.2010.01.011

Biron D, Shibuya M, Gabel C et al (2006) A diacylglycerol kinase modulates long-term thermotactic behavioral plasticity in *C. elegans*. Nat Neurosci 9:1499–1505. doi:10.1038/nn1796

Brenner S (1974) The genetics of *Caenorhabditis elegans*. Genetics 77:71–94

Chalfie M, Tu Y, Euskirchen G et al (1992) Green fluorescent protein as a marker for gene expression. Science 263:802–805

Chronis N, Zimmer M, Bargmann CI (2007) Microfluidics for in vivo imaging of neuronal and behavioral activity in *Caenorhabditis elegans*. Nat Methods 4:727–731. doi:10.1038/NMETH1075

Clark DA, Gabel CV, Gabel H, Samuel ADT (2007) Temporal activity patterns in thermosensory neurons of freely moving *Caenorhabditis elegans* encode spatial thermal gradients. J Neurosci 27:6083–6090. doi:10.1523/JNEUROSCI.1032-07.2007

Ezcurra M, Tanizawa Y, Swoboda P, Schafer WR (2011) Food sensitizes *C. elegans* avoidance behaviours through acute dopamine signalling. EMBO J 30:1110–1122. doi:10.1038/emboj.2011.22

Fang-Yen C, Wyart M, Xie J et al (2010) Biomechanical analysis of gait adaptation in the nematode *Caenorhabditis elegans*. Proc Natl Acad Sci USA 107:20323–20328. doi:10.1073/pnas.1003016107

Harris TW, Antoshechkin I, Bieri T et al (2010) WormBase: a comprehensive resource for nematode research. Nucleic Acids Res 38:D463–D467. doi:10.1093/nar/gkp952

Hedgecock EM, Russell RL (1975) Normal and mutant thermotaxis in the nematode *Caenorhabditis elegans*. Proc Natl Acad Sci USA 72:4061

Hoshi K, Shingai R (2006) Computer-driven automatic identification of locomotion states in *Caenorhabditis elegans*. J Neurosci Methods 157:355–363. doi:10.1016/j.jneumeth.2006.05.002

Hunt-newbury R, Viveiros R, Johnsen R et al (2007) High-throughput in vivo analysis of gene expression in *Caenorhabditis elegans*. PLoS Biol. doi:10.1371/journal.pbio.0050237

Ito H, Inada H, Mori I (2006) Quantitative analysis of thermotaxis in the nematode *Caenorhabditis elegans*. J Neurosci Methods 154:45–52. doi:10.1016/j.jneumeth.2005.11.011

Jurado P, Kodama E, Tanizawa Y, Mori I (2010) Distinct thermal migration behaviors in response to different thermal gradients in *Caenorhabditis elegans*. Genes Brain Behav 9:120–127. doi:10.1111/j.1601-183X.2009.00549.x

Kawano T, Po MD, Gao S et al (2011) An imbalancing act: gap junctions reduce the backward motor circuit activity to bias *C. elegans* for forward locomotion. Neuron 72:572–586. doi:10.1016/j.neuron.2011.09.005

Kimura KD, Miyawaki A, Matsumoto K, Mori I (2004) The *C. elegans* thermosensory neuron AFD responds to warming. Curr Biol 14:1291–1295. doi:10.1016/j

Kuhara A, Ohnishi N, Shimowada T, Mori I (2011) Neural coding in a single sensory neuron controlling opposite seeking behaviours in *Caenorhabditis elegans*. Nat Commun 2:355. doi:10.1038/ncomms1352

Lindsay TH, Thiele TR, Lockery SR (2011) Optogenetic analysis of synaptic transmission in the central nervous system of the nematode *Caenorhabditis elegans*. Nat Commun 2:306. doi:10.1038/ncomms1304

Miyara A, Ohta A, Okochi Y et al (2011) Novel and conserved protein macoilin is required for diverse neuronal functions in *Caenorhabditis elegans*. PLoS Genet 7:e1001384. doi:10.1371/journal.pgen.1001384

Miyawaki A, Llopis J, Heim R et al (1997) Fluorescent indicators for Ca^{2+} based on green fluorescent proteins and calmodulin. Nature 388:882–887

Mori I (1999) Genetics of chemotaxis and thermotaxis in the nematode *Caenorhabditis elegans*. Annu Rev Genet 33:399–422

Mori I, Ohshima Y (1995) Neural regulation of thermotaxis in *Caenorhabditis elegans*. Nature 376:344–348

Nagai T, Yamada S, Tominaga T et al (2004) Expanded dynamic range of fluorescent indicators for Ca(2+) by circularly permuted yellow fluorescent proteins. Proc Natl Acad Sci USA 101:10554–10559. doi:10.1073/pnas.0400417101

Nakai J, Ohkura M, Imoto K (2001) A high signal-to-noise Ca(2+) probe composed of a single green fluorescent protein. Nat Biotechnol 19:137–141. doi:10.1038/84397

Narayan A, Laurent G, Sternberg PW (2011) Transfer characteristics of a thermosensory synapse in *Caenorhabditis elegans*. Proc Natl Acad Sci USA 108:9667–9672. doi:10.1073/pnas.1106617108

Nishida Y, Sugi T, Nonomura M, Mori I (2011) Identification of the AFD neuron as the site of action of the CREB protein in *Caenorhabditis elegans* thermotaxis. EMBO Rep 12:855–862. doi:10.1038/embor.2011.120

Ohnishi N, Kuhara A, Nakamura F et al (2011) Bidirectional regulation of thermotaxis by glutamate transmissions in *Caenorhabditis elegans*. EMBO J 30:1376–1388. doi:10.1038/emboj.2011.13

Pierce-Shimomura JT, Chen BL, Mun JJ et al (2008) Genetic analysis of crawling and swimming locomotory patterns in *C. elegans*. Proc Natl Acad Sci USA 105:20982–20987. doi:10.1073/pnas.0810359105

Piggott BJ, Liu J, Feng Z et al (2011) The neural circuits and synaptic mechanisms underlying motor initiation in *C. elegans*. Cell 147:922–933. doi:10.1016/j.cell.2011.08.053

Ramot D, Johnson BE, Berry TL et al (2008a) The parallel worm tracker: a platform for measuring average speed and drug-induced paralysis in nematodes. PLoS One 3:e2208. doi:10.1371/journal.pone.0002208

Ramot D, MacInnis BL, Goodman MB (2008b) Bidirectional temperature-sensing by a single thermosensory neuron in *C. elegans*. Nat Neurosci 11:908–915. doi:10.1038/nn.2157

Sugi T, Nishida Y, Mori I (2011) Regulation of behavioral plasticity by systemic temperature signaling in *Caenorhabditis elegans*. Nat Neurosci 14:984–992. doi:10.1038/nn.2854

Tian L, Hires AS, Mao T et al (2009) Imaging neural activity in worms, flies and mice with improved GCaMP calcium indicators. Nat Methods 6:875–881

Yizhar O, Fenno LE, Davidson TJ et al (2011) Primer optogenetics in neural systems. Neuron 71:9–34. doi:10.1016/j.neuron.2011.06.004

Chapter 2
Classical Conditioning of the Proboscis Extension Reflex in the Honeybee

Yukihisa Matsumoto, Jean-Christophe Sandoz, and Martin Giurfa

Abstract Insects have sophisticated learning abilities subtended by simple neural systems and lower numbers of neurons compared to vertebrates. Especially, honeybees (*Apis mellifera*) are reported to have the highest and broad range of learning abilities. In this chapter, we introduce a classic behavioral tool for the study of olfactory learning and memory in bees, the olfactory classical conditioning of the proboscis extension reflex (PER). In this protocol, individually harnessed honeybees are trained to associate an odor with sucrose solution. The resulting olfactory learning is fast and induces robust olfactory memories that have been characterized at the behavioral, neuronal, and molecular levels. We detail step-by-step the methodology of olfactory PER conditioning in order to provide a standardized framework for experiments using this tool. We also review research highlights revealed by olfactory conditioning of PER and variations of this procedure applied in the case of honeybees.

Y. Matsumoto (✉)
Research Center for Animal Cognition, CNRS, University Paul Sabatier,
118 Route de Narbonne, 31062 Toulouse cedex 9, France

Graduate School of Life Science, Hokkaido University, Kita 10 Nishi 8,
Kita-ku, Sapporo 060-0819, Japan

Liberal Arts Science, Tokyo Medical and Dental University, 2-8-30 Kohnodai,
Ichikawa 272-0827, Japan
e-mail: yukihisa.las@tmd.ac.jp

J.-C. Sandoz
Research Center for Animal Cognition, CNRS, University Paul Sabatier,
118 Route de Narbonne, 31062 Toulouse cedex 9, France

Laboratory of Evolution, Genomes and Speciation, CNRS, 1 Avenue de la Terrasse,
91198 Gif-sur-Yvette, France

M. Giurfa
Research Center for Animal Cognition, CNRS, University Paul Sabatier,
118 Route de Narbonne, 31062 Toulouse cedex 9, France

Keywords Absolute conditioning • Classical conditioning • Differential conditioning • Honeybee • Learning • Memory • Olfactory conditioning • Proboscis extension reflex (PER)

2.1 Introduction

Associative learning is a fundamental property of nervous systems, which is governed by conserved rules both across species and across modalities. In a classical conditioning procedure, the animal is presented with two types of stimuli, the unconditioned stimulus (US) and the conditioned stimulus (CS). The US is a stimulus which innately evokes a certain response, while the CS is usually a neutral stimulus without any initial connection with a response. Through forward pairing of CS and US, the animal learns that the CS predicts US delivery and starts responding to the CS. The response evoked by the CS is termed conditioned response. Classical conditioning is also called "Pavlovian conditioning," after the work by Ivan Pavlov who laid the foundations of this conditioning by studying conditioned salivation in dogs, which resulted from pairings between sounds (CS) and meat (US) (Pavlov 1927).

Insects have good learning abilities subtended by a more simple neural system and lower numbers of neurons than vertebrates (Mizunami et al. 2004; Giurfa 2007). For this reason, several species of insects have become mainstream models for research on learning and memory (Mizunami et al. 2004). Thus, insect species as diverse as honeybees, fruit flies, crickets, cockroaches, ants, and moths have been shown to possess robust learning abilities, using behavioral experiments (Bitterman et al. 1983; Tully and Quinn 1985; Matsumoto and Mizunami 2000, 2002; Balderrama 1980; Dupuy et al. 2006; Daly and Smith 2000). Among these insects, honeybees (*Apis mellifera*) are reported to have the highest and broad range of learning abilities (e.g., Menzel 1999; Giurfa 2007; Avarguès-Weber et al. 2011; Sandoz 2011).

Honeybees are able to associate odors, colors, visual patterns, or tactile stimuli with a food reward (Menzel et al. 1993; Menzel 1999; Giurfa 2007). However, studies on learning and memory in honeybees have mostly used visual learning protocols when the focus was on the performances of free-flying honeybees and olfactory learning protocols when the goal was a full control of behavior in harnessed bees. The latter are based on the conditioning of the proboscis extension reflex (PER), a case of classical conditioning which is relatively easy to carry out in the laboratory (Giurfa 2007; Giurfa and Sandoz 2012).

The PER is a reflexive response which is part of the bee's feeding behavior while foraging or within the hive (Frings 1944). It is observed when the antennae, tarsi, or mouthparts of a hungry honeybee come in contact with sugar solution; the bee then automatically extends its proboscis to reach the sucrose solution (PER) and drink it. In naïve bees, odors generally do not evoke the PER. During conditioning, an odor (CS) is presented in close temporal association with a sucrose solution reward (US).

At the end of training, the odor alone elicits the PER, indicating that the bee has learnt the odor-reward association (Takeda 1961; Bitterman et al. 1983). The PER is usually recorded as a dichotomous response (1 or 0), which can thus be used as a precise index for learning and memory performance.

Olfactory conditioning of the PER is performed at an individual level, on immobilized honeybees placed in small harness tubes. This allows simultaneous or consecutive monitoring of behavior and neuronal activity, for instance, using neurophysiological or optophysiological techniques, during memory acquisition and memory retrieval (e.g., Faber et al. 1999; Okada et al. 2007; Strube-Bloss et al. 2011). In this chapter, we detail a simple protocol for absolute classical conditioning of the PER which can be established with minimal investment in any laboratory.

2.2 Preparation for the Experiment

2.2.1 *Materials*

Honeybees, 20 harness tubes, tube rack, ice water, 30 mL glass screw vial (make small vent holes on the cap to avoid suffocation), 50 % (weight/weight) sucrose solution (1.80 M), toothpick, odorants, filter papers, 20 mL syringes, and latex examination gloves.

2.2.2 *Honeybees*

When adult worker bees age, their tasks shift from nursing to guarding and then to foraging (von Frisch 1967; Seeley 1982, 1995). Foragers, the older bees that go out of the hive to collect food, are reported to have the highest learning ability among bees working on other tasks (nurse, guardian) (Ray and Ferneyhough 1999). This makes sense as the subjective value of sucrose reward is probably higher in animals whose foraging motivation is also higher. This is why a control of the kind of bees used in olfactory PER conditioning experiments is desirable. Confounding several casts of bees in the same set of data will mask specific learning abilities and pool together animals that differ dramatically in their unconditioned response to sucrose reward (Scheiner et al. 2001). Generally, foragers are captured while leaving the hive and used for experiments (e.g., Guerrieri et al. 2005a). In other cases, emerging bees are placed in cages in an incubator and maintained in controlled conditions until reaching foraging age (Pham-Delègue et al. 1993). Even more recommendable, in order to ensure the presence of foragers in the bees to be conditioned, is the previous training to an ad libitum sucrose feeder placed close to the hive on which regular foragers will be collected for experiments.

2.2.3 Pyramid for Catching Bees

A pyramid (height 24 cm, apex 3.5 × 3.5 cm, base 18 × 18 cm) made of UV-translucent Plexiglas is useful to catch bees when leaving the hive and flying towards the sky. This is why the Plexiglas has to be UV transparent, in order to offer a complete view of celestial cues and thus lure the departing bees within the pyramid. The pyramid is closable at the apex and at the base. For catching bees, the pyramid is held at about 10–20 cm distance in front of the hive entrance with the base open and the apex closed. Bees will naturally enter the pyramid when leaving the hive. When enough bees have been caught, the base should be closed and the pyramid taken to the laboratory. When the pyramid is darkened (except for its apex), bees will exit it through the apex because of their positive phototaxis. Given its pyramidal form, the bees will tend to leave the pyramid one by one, thus facilitating individual capture and precluding mass escape. This allows transferring the bees from the pyramid apex into glass vials.

2.2.4 Harness Tubes and Tube Rack

Harness tubes (major diameter 10 mm, height 32 mm) can be made by cutting cylinders. Materials can be metal, plastic, acryl, etc., as far as it suits the experiment. A rack with numbered boreholes would be useful for handling and identification of harnessed bees. Each harness tube should be numbered to allow individual identification of the bees throughout the experiment. When placed in the tubes, only the bee head should protrude, thus hiding other body parts from possible contacts with later sucrose stimulation. The forelegs of the bees, for instance, should not be able to move free but should remain enclosed without the tube in order to avoid uncontrolled cleaning of mouth pieces and antennae during the experiment which may interfere with olfactory and sucrose stimulation.

2.2.5 Odorants

The choice of odorants should be made according to the main purpose of the experiment. Both single chemical compounds (alcohols, aldehydes, terpenes, etc.) and mixtures of compounds (rose extract, carnation extract) have been used in such experiments (see Sandoz 2011). It should be noted that the use of some odors as conditioned stimuli, like pheromones, which are potentially important in the bee's biology, may induce a potential bias when used in experiments on learning and memory. Likewise, possible prior experience of bees with some odorants (in the hive or while foraging) may also affect the results. Such responses indicate that the bees may have already learned this odorant in an appetitive context before.

One way to get rid of this effect is to discard those bees that exhibit spontaneous PER upon the first olfactory stimulation (first conditioning trial).

We recommend always using two different odors in any experiment in which an absolute-conditioning protocol will be used. In absolute conditioning animals learn that a unique CS is reinforced. Yet, to show that the acquisition evinced in a learning experiment is stimulus specific (here odor specific), it is recommendable to run the experiments with two parallel groups, one group trained with a CS1 paired with the US and another group trained with a CS2 paired with the US. Using two odors has the additional advantage of allowing a test of memory specificity: each bee is trained with one odor (say CS1) and will be tested afterwards with two odors, the conditioned odor (CS1) and the other odor (the CS2 used for the other group) which will be novel to it (NO). In this way, it is possible to distinguish between memories that are odor specific (CS specific), and which should be only evoked by the CS1, and unspecific PER responses which are elicited by the NO. The difference between these two response categories (to the CS and to the NO) provides an assessment of the CS-specific memory. As mentioned above, to balance the effect of the two chosen odorants in the experiments, half the bees should be conditioned with CS1 and the other half with CS2. The use of two odors that are well distinguished from each other and are easily associated with sucrose solution is recommended (e.g., 1-nonanol and 2-hexanol) (Guerrieri et al. 2005a).

2.2.6 Odor Ventilation

When carrying out olfactory learning experiments, the odors presented should be exhausted from the experimental system as soon as the stimulation ends. This ensures that the temporal properties of the stimulus are well controlled. Therefore, a ventilation system should be made with duct hose, one end being set at the experimental setup and the other end connected to a standard air extraction. During conditioning and test trials, honeybees are set individually in front of the odor stimulation device, with the exhaust at their back. Ventilation should not be too strong, as unintended mechanical stimulations from the ventilation may interfere with olfactory learning.

2.2.7 Olfactory Stimulation (CS)

In the laboratory, researchers use computer-controlled odor stimulation devices (Galizia et al. 1997) to deliver the CS to antennae of the conditioned bees. Although such systems allow well-controlled olfactory stimulation, PER conditioning can be also performed using simple plastic syringes. With gloves on, take two 20 mL syringes and set a piece of filter paper (10 × 30 mm) inside each of them. Soak one of the filter papers with 5 μL of 2-hexanol (syringe 1) and the other with 5 μL of 1-nonanol (syringe 2). The syringe method underlines again the importance of using two odors in any absolute-conditioning experiment. The mechanic stimulation of the air puff could act as a confounding CS in conditioning trials. The test with the novel odor is therefore of fundamental importance to verify that such mechanic stimulation is not driving the bee's responses, in which case PER will be specific to the odor learned.

2.2.8 Sucrose Solution (Reward US)

For US sucrose solution (usually 50 %, weight/weight, i.e., 1.80 M), dissolve sucrose in a vial of distilled water by 50 % weight/weight. The choice of sucrose concentration is crucial for the experimental success. Indeed, diluted sucrose concentrations (below 20 %) are suboptimal for the appetitive motivation of the bees, while highly concentrated solutions are also suboptimal due to their high viscosity, which renders difficult ingestion through the proboscis (Farina and Núñez 1991). Thus, concentration in the range of 30–50 % should be used. Dispense 1 mL sucrose solution into Eppendorf tubes and keep them in the freezer until needed. Defrost a tube to room temperature before each experiment.

2.3 Protocol for Absolute Classical Conditioning

2.3.1 Catching and Harnessing Bees in Harness Tubes

1. Capture honeybee foragers outside the hives with a pyramid (see above, or a sweep net if not available) in the morning of the day of the experiment or in the afternoon of the day before the experiment.
2. Put one to five honeybees in a screw vial immediately after capture, and cool the vial in ice water for about 3 min to anesthetize the bees. Cooling time should be kept to a minimum as extended cooling could impair learning performances (e.g., Frost et al. 2011). As soon as the bees are immobilized, place each bee in a harness tube using a piece of tape behind the head. In this position, mouthparts and antennae should be able to move freely. Take care of not leaving the forelegs free to move and reach the head once the bee has been harnessed.

How many bees have to be used per experiment is a critical question as one ideally wants large sample sizes that ensure statistical power. Low sample sizes (e.g., around ten individuals per group) have to be avoided; if possible, use 40–50 bees per experiment (see below, Data Analysis).

2 Classical Conditioning of the PER

Fig. 2.1 Proboscis extension reflex (PER) in harnessed honeybee. Contact of the antennae and/or proboscis with sucrose solution elicits the PER. Honeybees are conditioned to extend their proboscis in response to an odor (CS) when the odor is presented contingent upon a sucrose stimulus (US)

2.3.2 Feeding Bees

Depending on the interval between capture and the start of the conditioning, it can be necessary to feed the bees with 50 % sucrose solution: if bees were caught in the morning of the experimental day and conditioning starts after 3–4 h, then feeding is unnecessary; if, however, conditioning starts in the afternoon, feed a drop (~5 µL) of sucrose solution approximately 30 min after fixation to avoid excessive starvation and resulting mortality before start of the experiment; lastly, if bees were caught on the day before the experiment, they should be fed to satiation and kept in a dark, humid container at room temperature (20–25 °C) overnight. It should be kept in mind that these are general rules of thumb, which should be adapted depending on the season, local conditions, etc. Feeding will reduce appetitive motivation and thus unconditioned responses (PER) to sucrose. Therefore, a good balance should be kept between starvation and feeding, to keep bees with good appetitive motivation and sufficient vitality.

Use a toothpick to present sucrose solution to a bee or a micropipette with automatized volume delivery if you want to control this factor. When sucrose solution touches the antennae, a hungry bee will elicit PER (Fig. 2.1). Let the bee lick the solution by touching the proboscis with the toothpick. Bees that do not show PER are either satiated or in poor physical condition. Such bees should be removed from the experiment due to the lack of unconditioned responses.

2.3.3 Absolute Conditioning

As mentioned above, absolute conditioning is the simplest form of associative conditioning (paired conditioning), in which a single stimulus (odor, CS) is reinforced

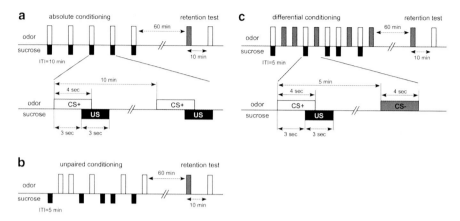

Fig. 2.2 Experimental procedures of three types of olfactory classical conditioning using PER. (**a**) Absolute conditioning (paired conditioning). Absolute conditioning is a pairing of 4 s of an odor (CS+: *white bars*) and subsequent 3 s of sucrose solution (US: *black squares*) with 1 s overlap. Here, animals receive five trials of paired conditionings with intertrial interval of 10 min and are tested with two odors, conditioned odor (CS: *white bar*) and novel odor (NO: *gray bar*), 60 min after conditioning. (**b**) Unpaired conditioning. In unpaired conditioning group, animals receive explicitly unpaired presentations of the CS and of the US (five odor-only and five sucrose-only presentations, 5 min apart in a pseudorandomized sequence) and are tested with CS (*white bar*) and NO (*gray bar*). (**c**) Differential conditioning. During differential conditioning, one odor is paired with sucrose (CS+: *white bars*), and the other odor is presented without sucrose (CS−: *gray bars*)

(sucrose solution, US). Here, to observe trained-odor-specific memory, we will use one odor (CS1) for conditioning and two odors (CS1 and NO, 1-nonanol and 2-hexanol) in memory retention tests. Parameters such as the number of trials and the duration of the intertrial interval (ITI) can be manipulated according to the purpose of the experiment. In this text, we will describe the standard procedure of absolute conditioning with 5 CS–US pairings with intertrial intervals of 10 min (Fig. 2.2a). This procedure yields a robust and stable long-term memory that can be retrieved several days after conditioning (>4 days) and that is protein synthesis dependent (Menzel 1999). Learning efficiency is highly affected by circadian rhythm and is lowest around early evening (Lehmann et al. 2011). Thus, conditioning should be planned to end before 1600 hours. Twenty bees are used in this experimental procedure.

1. Place two syringes for olfactory stimulation, sucrose solution (in an Eppendorf tube), and a toothpick on the table. If using an odor stimulation device, this should be checked before beginning.
2. Set the first honeybee (bee A) in front of the exhaust. The bee should be left to familiarize with the experimental situation for at least 15 s before applying any stimulation.

3. Deliver the CS and the US in an appropriate temporal relationship. Standard stimulation consists of a pairing of 4 s of odor (CS1) and subsequent 3 s of sucrose solution (US) with an interstimulus interval of 3 s (1 s overlap). Record the presence (+) or absence (−) of PER in the 3 s between odor onset and sucrose delivery (conditioned response to the odor). If the bee responds with a PER to the odor, touch the proboscis with sucrose solution and let the bee drink for 3 s. If it did not show PER to the odor, touch the antennae with the toothpick first to elicit PER and present sucrose solution to the bee's extended proboscis with the toothpick. Let the bee drink for 3 s.
4. Leave the bee in the conditioning place for 10 s after the CS–US pairing. This is important for the contextual cues around the setup to lose an anticipating, predictive link to the US. Place the next subject (bee B), and start (2–4) again.
5. Perform the first conditioning trial in all 20 subjects, recording each time PER and sucrose solution intake, by repeating (2–4) (first-trial conditioning).
6. Repeat (2–4) four more times (five trials per bee in total) with an ITI of 10 min. If you use 20 bees and exchange the subject every 30 s, which corresponds to the duration of a conditioning trial, the second trial of the first bee (bee A) will just follow the first trial of the 20th bee (bee T), and the ITI between first and second trial will be 10 min as planned.

2.3.4 Keeping Bees Between Training and Retention Tests

Depending on the experiment and thus on the question raised, bees may need to be tested to assess the presence of memory from a few minutes to several days after training. If you are planning to take several hours between conditioning trials and memory retention test, keep the bees in a dark, humid container at room temperature until the test to maintain their physical condition.

Bees that have to be tested several days after conditioning will be subjected to different handling procedures to ensure survival until the respective retention test. Bees tested 1 day after conditioning can be kept in the harness tubes because mortality is usually low in these conditions; yet, bees should be fed to satiation with 50 % sucrose solution, at least 60 min after the end of conditioning to ensure survival. Bees tested more than 2 days after conditioning should be kept in small cages because mortality will increase in prolonged harnessing conditions. To this end, bees need to be individually identified by means of color marks painted on the thorax with watercolors, following a code that allows later recognition. Bees are then placed in groups of ~30 individuals in small cages (e.g., 65 × 70 × 25 mm) supplied with water and a diet of 50 % sucrose and 50 % honey mixture ad libitum. The cages should be kept in a dark and humid container at room temperature. On the morning of the test day, bees are transferred from the cages into glass vials, cooled on ice, and placed again individually in the harness tubes. Retrieval tests are usually performed after 5 h of food deprivation to ensure adequate appetitive motivation.

2.3.5 Memory Retention Test

Memory retention tests are performed to test for PER to two odors (CS, the conditioned odor, and NO, the novel odor) delivered by two different odor syringes. The order of presentations of the two odors should be randomized between bees to avoid sequential effects. Thus, ten subjects (bee A–J) are first tested with CS1 first and NO second, while the next ten subjects (bee K–T) are tested with the reversed odor of presentation. As for conditioning, the ITI between odor stimulations is 10 min. Thus, proceeding like during the conditioning trials (see above) but without US delivery is recommended.

2.3.6 Response Check for Sucrose Solution

At last, i.e., following the last retention test, the unconditioned response (PER to sucrose) should be tested in all animals by applying sucrose solution to the antennae. Bees that do not show the unconditioned response should be discarded from the data, as their lack of response to the odors cannot be necessarily ascribed to a lack of memory but could be due to a low physical condition.

2.3.7 Unpaired Conditioning

To ensure that honeybees acquired associative memory by absolute conditioning (paired conditioning), several kinds of experimental controls can be performed in parallel to normal conditioning. One of them is the explicitly unpaired conditioning that has to be performed with a number of bees equivalent to that used in the absolute-conditioning group. The results of both groups, ran in parallel, are compared to determine whether increases in conditioned responses in the absolute-conditioning group are the consequence of real associative learning. In the explicitly unpaired group, bees receive unpaired presentations of the CS and of the US (five odor-only and five sucrose-only presentations, 5 min apart in a pseudorandomized sequence; Fig. 2.2b). Thus, both the absolute-conditioning (the paired group) and the unpaired group have exactly the same sensory experience (five CS and five US presentations), the difference being in the pairing or absence of pairing between odor and sucrose. An additional factor that needs to be controlled is the fact that the number of placements in the setup is twice as higher (if not controlled) in the unpaired group compared to the absolute-conditioning group. This can be balanced by inserting between conditioning trials five blank trials in the absolute-conditioning group, in which the bee will be simply located in the setup without stimulation.

As responses to sucrose solution and learning performance may vary according to the season and climate conditions (Ray and Ferneyhough 1997a, b), the same number of experimental and control animals should always be conditioned every day.

2.3.8 Differential Classical Conditioning

Differential conditioning is an associative conditioning procedure in which animals have to learn the difference between two or more stimuli, based on their differential outcome in terms of reinforcement. In the case of olfactory PER conditioning, differential conditioning is achieved by using two odors, one (the CS+) that is paired with sucrose solution and the other (the CS−) that is presented explicitly without reinforcement (Fig. 2.2c). Compared to the absolute-conditioning protocol, differential conditioning has the advantage of providing the researcher with an internal control of the associative nature of the established memory, as bees have to learn to respond to the CS+ but not to the CS−. Yet its use depends on the question raised by the experimenter. Note, for instance, that differential conditioning may boost discrimination capabilities as in the former case, bees are trained to discriminate stimuli while in the later case, they are trained to respond to a unique stimulus. Perceptual measures derived from one protocol or the other may, therefore, differ.

2.3.9 Data Collection and Making Graphs

For each conditioning trial, the percentage of conditioned responses (%CR) is calculated as the number of bees showing PER to the conditioned odor with respect to the total number of bees assayed. Typical results of absolute olfactory classical conditioning of PER are shown in Fig. 2.3a–c. In Fig 2.3a, plotted CR (%) during conditioning trials corresponds to an acquisition curve. Given the dichotomous nature of the response measured (PER: 1 or 0), it is obvious that the gradually changing acquisition curve does not necessarily reflect the stepwise nature of individual bee responses; yet it provides a basis to assess populational learning and to promote analyses of individual performances with respect of group ones (Pamir et al. 2011). In Fig. 2.3a, we see that bees from the paired group increased their responses to the CS, while bees from the unpaired group do not show such increase, thus showing the associative nature of performance variation in the paired group. For memory retention tests, the percentage of PER both to the learned odor (CS) and to the novel odor (NO) has to be plotted (Fig. 2.3b). In the tests after 60 min, which corresponds to a midterm memory (MTM), paired bees responded to the CS but not to the NO, thus showing the presence of CS-specific MTM (Fig. 2.3b). To give a more precise measure of odor-specific memory, the percentage of bees that showed PER to the CS+ but not to the NO can be plotted (specific response: %SR) (Fig. 2.3c). Figure 2.3d,e show typical results of a differential-conditioning procedure. Bees increased their responses to the CS+ but not to the CS− (Fig. 2.3d). Likewise, in the 60 min retention tests, bees responded to the CS+ and not to the CS− (Fig. 2.3e).

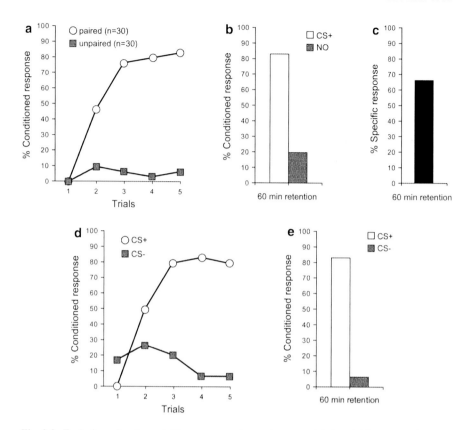

Fig. 2.3 Typical results of acquisition curves and retention tests. (**a**) Acquisition curves for honeybees trained by paired conditioning (*white circle*) or unpaired conditioning (*gray square*). For each conditioning trial, conditioned response (%CR) is calculated as percentage of bees that showed PER to the conditioned odor. (**b, c**) Memory retention tests for honeybees trained by paired conditioning. For the test 60 min after conditioning, both CR for a conditioned odor (CS+: *white graph*) and CR for a novel odor (NO: *gray graph*) are plotted in (**b**), and specific response (%SR: *black graph*, see text) are calculated and plotted in (**c**). (**d**) Acquisition curves for honeybees trained by differential conditioning. (**e**) Memory retention tests for honeybees trained by differential conditioning

2.3.10 Data Analysis

To analyze acquisition within a single group, a Cochran Q test can be used as it is especially conceived for a repeated-measures experimental design with a dichotomous variable. To compare acquisition performances between two or more different groups (e.g., paired vs. unpaired groups; see above), Mann–Whitney U tests can be used on the sum of responses to the CS observed during conditioning. Yet, this method has the disadvantage of losing the dynamics of an acquisition curve as it reduces it to a single data point. A similar criticism applies to the use of Fisher exact tests and/or χ^2 tests to compare between groups the summed CS responses or the CS responses in the last acquisition trial. Obviously, one may want to focus just on the last acquisition trial and in this case, use of these tests is appropriate. However, if entire acquisition curves are to be compared between groups, other solutions have to be found. Cochran Q test cannot be used to this end as it is a within-group test. Thus, repeated-measures analyses applicable for between-group comparisons have to be used. A solution to this problem is the use of standard two-factor analysis of variance (ANOVA) for repeated measurements, with one factor being the treatment to be analyzed (e.g., paired vs. unpaired CS responses) and the other factor the response along trials (e.g., trials one to five in the example of Fig. 2.3a,d). The interaction between both factors will also be computed allowing the detection of specific treatment x trial effects. ANOVA procedures are in principle not applicable in the case of dichotomous data; yet Monte Carlo studies have shown that it is permissible to use ANOVA on dichotomous data (Lunney 1970) if comparisons imply equal cell frequencies and at least 40 degrees of freedom of the error term. By fulfilling these conditions, the use of repeated measurement ANOVA will allow not only between-group comparisons but also within-group analysis as achieved by the Cochran test.

To compare response levels to the CS and the novel odor (NO) in the retention tests, a McNemar test is typically used as it is applicable for paired-sample testing of dichotomous data. Note that given the dichotomous nature of responses, one has to consider not only correct responses (i.e., CS+:1/NO:0) but also the three possible cases of incorrect responses (i.e., CS+:1/NO:1; CS+:0/NO:0; CS+:0/NO:1).

2.4 Research on Honeybee Learning and Memory Using Olfactory Classical Conditioning of PER

In this chapter, we have introduced five-trial absolute conditioning with ITI of 10 min. The choice of such a long interval determines a conditioning with "spaced trials," in contrast with protocols employing "massed trials" in which trials are separated by very short (typically 1 min) intervals. Extensive study of associative olfactory conditioning of PER revealed the existence of different memories, depending on US intensity (i.e., the amount and/or quality of sucrose solution received during conditioning), the number of conditioning trials, and the ITI (Menzel 1999, 2001; Menzel et al. 2001). The memory induced by a single conditioning trial decays rather quickly, mostly after 1 day (Menzel 1990; Hammer and Menzel 1995) and is sensitive to amnestic treatments (Menzel et al. 1974; Erber et al. 1980). This memory can be dissected into two memory phases, short-term memory (STM) and medium-term memory (MTM). These phases are independent of protein synthesis (Grünbaum and Müller 1998). In contrast, multiple conditioning trials induce a stable, long-lasting memory, which can be retrieved several days after conditioning (Menzel et al. 2001). Trial spacing is a dominant factor for both acquisition and retention. Generally, massed trials lead to impaired memory performances compared to spaced trials. Disruption of molecular processes by pharmacological and genetic tools showed that memory formed by procedures using spaced trials is dissectible into STM and subsequently three independent, parallel phases: MTM, early long-term memory (eLTM), and late long-term memory (lLTM). MTM in the hours range requires constitutive PKC activity (Grünbaum and Müller 1998). Both eLTM and lLTM require PKA- and nitric oxide (NO)-dependent processes for their formation (Müller 1996, 2000). However, eLTM, retrievable 1–2 days after conditioning, requires translation whereas lLTM demands for transcription (Friedrich et al. 2004).

Recent experiments based on neuropharmacological manipulation or RNA interference of PER conditioning identified several molecular processes important for olfactory memory formation. Most importantly, these experiments demonstrate that different molecular processes subtend different memory phases. For example, glutamate and N-methyl-D-aspartate (NMDA) receptors are involved in MTM (Müßig et al. 2010) and eLTM formation but not in lLTM formation (Maleszka et al. 2000; Si et al. 2004; Locatelli et al. 2005; Müßig et al. 2010). Alpha-bungarotoxin (BGT)-sensitive nicotinic acetylcholine (nACh) receptors are involved in eLTM formation, while BGT-insensitive nACh receptors are involved in memory retrieval (Gauthier et al. 2006). Intracellular calcium, adenylyl cyclase, cyclic nucleotide-gated channels, calmodulin (CaM), and CaMKII are all involved in lLTM formation but not in eLTM formation (Perisse et al. 2009; Matsumoto et al. unpublished data). A crucial future challenge will be to clarify how these molecules interact for giving rise to the different olfactory memory phases subtending retention.

The neuronal circuits processing the odor stimulus (CS) and the sucrose reward (US) in PER conditioning are well described (Menzel 1999; Giurfa 2007; Giurfa and Sandoz 2012). The CS pathway includes the olfactory receptors located on the antennae, the antennal lobes (ALs: primary olfactory centers), the mushroom bodies (MBs: higher-order centers), and the lateral protocerebrum (premotor output regions). Olfactory information detected at the level of the antennae is processed in the ALs, which then send this information to the MBs input region (calyces) and to the lateral protocerebrum via projection-neuron tracts. The MBs, with their intrinsic Kenyon cells, process input from different sensory modalities (Mobbs 1982; Abel et al. 2001; Gronenberg 2001) and their extrinsic neurons are multimodal (Grünewald 1999; Mauelshagen 1993; Okada et al. 2007; Haehnel and Menzel 2010). Concerning the US pathway, information from sucrose receptors located on the antennae and the proboscis is relayed to the subesophageal ganglion. Directly or indirectly they contact the ventral unpaired median neuron number 1 of the maxillary neuromere (VUMmx1 neuron), which projects widely in the ALs, the calyces of the MBs, and the lateral protocerebrum. Activity of this individual neuron can substitute for the US in classical conditioning assays (Hammer 1993). The ALs, the MBs, and the lateral protocerebrum are thus main convergence sites for the CS and US pathways.

PER conditioning has allowed showing the existence of learning-dependent plasticity in the CS and US pathways. Changes in neural activity and in synaptic architecture have been reported by application of electrophysiological, optophysiological, and histological techniques (Hammer 1993; Mauelshagen 1993; Okada et al. 2007;

Szyszka et al. 2008; Hourcade et al. 2010). For example, plasticity of single identified neurons in the bee brain, such as the abovementioned VUMmx1 or of the MB-extrinsic neuron PE1, after olfactory PER conditioning was investigated by intracellular recordings coupled to PER conditioning (Hammer 1993; Mauelshagen 1993) or extracellular recordings (Okada et al. 2007). Neuroanatomical analyses have also been employed to assess the effect of olfactory memory formation at the structural level. It has been shown, for instance, that the density of local microcircuits (called microglomeruli) in the olfactory region of the MB calyces was increased upon olfactory lLTM formation (Hourcade et al. 2010).

On a behavioral level, the advent of olfactory classical conditioning of PER has allowed studying from a psychological perspective several learning-related phenomena well known in vertebrates, such as overshadowing (Smith 1998), blocking (Smith and Cobey 1994; Gerber and Ullrich 1999; Hosler and Smith 2000; Guerrieri et al. 2005b), second-order conditioning (Hussaini et al. 2007), sensory preconditioning (Müller et al. 2000), positive and negative patterning (Deisig et al. 2001, 2002, 2003), spontaneous recovery from extinction (Sandoz and Pham-Delègue 2004; Stollhoff et al. 2005), reversal learning (Komischke et al. 2002; Devaud et al. 2007; Hadar and Menzel 2010), and reconsolidation (Stollhoff et al. 2008) in a controlled way.

2.5 Other Forms of PER Conditioning in Honeybees

Researchers manipulated classical conditioning of PER to associate reward with other sensory stimuli such as monochromatic lights (Hori et al. 2006), motion cues (Hori et al. 2007), polarized light (Sakura et al. 2012), antennal mechanosensory stimulations (Giurfa and Malun 2004), or antennal temperature stimulations (Hammer et al. 2009). These are appetitive learning paradigms that use sucrose solution as positive reinforcer. Besides this, aversive learning paradigms have also been developed in bees, using mild electric shocks as negative reinforcer (Vergoz et al. 2007), yet in this case PER is no longer considered but the sting extension reflex (SER). In this aversive learning protocol, an odor CS is associated with an electric shock US, and the bees learn to respond to the trained odor with an extension of their sting. Pharmacological approaches using olfactory appetitive conditioning of PER and olfactory aversive conditioning of SER suggested that octopamine and dopamine subserve appetitive and aversive reinforcement in the honeybee, respectively (Hammer and Menzel 1998; Vergoz et al. 2007).

Recently, Ayestaran et al. (2010) showed by means of olfactory conditioning of PER that aversive substances, such as quinine, salicine, and amygdalin, can induce a post-ingestional malaise that will later reduce the bees' tendency to respond to the conditioned odor. Wright et al. (2010) also associated odors with toxins, namely, quinine or amygdalin, mixed in sucrose solution. This work demonstrated that two distinct monoaminergic pathways, mediated by dopamine and serotonin, respectively, account for conditioned food aversion in honeybees: learned avoidance of bitter substances is primarily modulated by dopamine, while learning to associate odors with the malaise caused by ingesting amygdalin is mediated by serotonin. The first hypothesis is under debate as in other experiments with harnessed bees avoidance of bitter substances could not be clearly observed (de Brito sanchez et al. 2005; Ayestaran et al. 2010).

Apart from the learning paradigms on harnessed bees, we have described, until now, the study of honeybee perception and learning started with visual learning paradigms on free-flying bees. A long experience in such protocols has shown that free-flying honeybees can be conditioned to associate a plethora of sensory stimuli with sucrose reward, like colors, shapes and patterns, depth, and motion contrast, among others (von Frisch 1914; Wehner 1981; Giurfa and Menzel 1997; Lehrer 1997; Giurfa and Lehrer 2001; see Avarguès-Weber et al. 2011 for review). This experimental paradigm demonstrated that honeybees learn to categorize visual stimuli based on perceptual similarity (Giurfa et al. 1996; Stach et al. 2004) and also to classify stimuli based on conceptual rules such as "same," "different," "above," and "below" (Giurfa et al. 2001; Avarguès-Weber et al. 2011), thus showing the presence of higher-order learning capacities in honeybees.

2.6 Conclusion

Honeybee provides a model system for the study of neural and molecular substrates of learning and memory and basic cognitive faculties. Olfactory conditioning of PER in harnessed bees is a particularly helpful protocol to this end as it allows simultaneous or consecutive monitoring of behavior and of neuronal activity using extra- and intracellular recording or optical imaging of neural activity (e.g., Hammer 1993; Faber et al. 1999; Okada et al. 2007; Denker et al. 2010). Local injection or uncaging of neuromodulatory compounds enables studying the molecular basis underlying memory formation at local brain region level (Müller 2000; Devaud et al. 2007; Perisse et al. 2009). Moreover, a great variety of learning paradigms with different CSs and USs can be applied and may help uncovering whether molecular processes underlying olfactory PER conditioning are general phenomena shared by other paradigms. The genome sequence of the honeybee *A. mellifera* has been made available (The Honeybee Genome Sequencing Consortium 2006), and application of molecular biological techniques such as RNA interference to investigate honeybee learning and memory has already begun. If new molecular genetics techniques allowing expression or blocking of the expression of particular genes in specific regions of the honeybee brain appear, this animal model may yet provide novel breakthroughs in the study of the neural basis of learning, memory, and cognition.

Here we have aimed at providing a didactic and detailed explanation about basic procedures to be followed when performing olfactory PER conditioning. We hope, in this way, that researchers who are unfamiliar (or partially familiar) with this protocol will become attracted to it by its easiness and robustness and will therefore help increasing the efforts towards the novel breakthroughs mentioned above.

Appendix A Column: History of a Classical Conditioning of PER in Honeybees

Long before research on PER conditioning started, it was well known that the PER could be elicited by stimulating gustatory organs like the antennae, tarsi, or mouthparts with sugar solution. The PER had thus been detected in bees (Frings 1944; Frings and Frings 1949), flies (Minnich 1926), and butterflies (Minnich 1921), among others. Later, a Japanese researcher who had worked with Karl von Frisch, Masutaro Kuwabara, realized that this appetitive reflex could be conditioned using visual stimuli as CS and sucrose solution delivered to the tarsi as US (Kuwabara 1957). However, Kuwabara's work did not receive broad attention as shown by the fact that almost 50 years had to pass before other researchers published results on honeybee visual conditioning using Kuwabara's method (Hori et al. 2006; Hori et al. 2007). For this conditioning to work, Kuwabara and Hori et al. had to cut the bees' antennae. The low acquisition rates observed in antennae-deprived bees despite long conditioning procedures (Hori et al. 2006; Hori et al. 2007) may be related to this fact. It has been recently shown that antennae deprivation reduces sucrose responsiveness when measured through tarsal stimulation (de Brito Sanchez et al. 2008), probably leading to a reduction of US value and of acquisition and retention performances.

The olfactory conditioning of PER was afterwards established by a student of Kuwabara, Kimihisa Takeda, who reported on this procedure in 1961 (Takeda 1961) using odors as CS and sucrose solution as US. As was common use 50 years ago, Takeda did not report any acquisition, retention, or extinction curves, nor did he provide any statistical analysis of PER responses. He only presented tables with the raw data of single bees, with very low sample sizes. Despite data paucity, lack of statistics, absence of controls, and representative sample sizes, Takeda's work laid down the experimental principles and the scientific questions that would serve as a basis for future, more controlled research on honeybee learning and memory. He showed for the first time extinction learning (including spontaneous recovery), stimulus generalization and discrimination, conditioned inhibition, and second-order conditioning in the olfactory domain in honeybees. In this way he established olfactory PER conditioning as a useful tool for the study of invertebrate learning and memory.

References

Abel R, Rybak J, Menzel R (2001) Structure and response patterns of olfactory interneurons in the honeybee, *Apis mellifera*. J Comp Neurol 437:363–383

Avarguès-Weber A, Deisig N, Giurfa M (2011) Visual cognition in social insects. Annu Rev Entomol 56:423–443

Ayestaran A, Giurfa M, de Bitro Sanchez MG (2010) Toxic but drank: gustatory aversive compounds induce post-ingestional malaise in harnessed honeybees. PLoS One 5:e15000. doi:10.1371/journal.pone.0015000

Balderrama N (1980) One trial learning in the American cockroach, *Periplaneta americana*. J Insect Physiol 26:499–504

Bitterman ME, Menzel R, Fietz A, Schäfer S (1983) Classical conditioning of proboscis extension in honeybees (*Apis mellifera*). J Comp Psychol 97:107–119

Daly KC, Smith BH (2000) Associative olfactory learning in the moth *Manduca sexta*. J Exp Biol 203:2025–2038

de Brito Sanchez MG, Giurfa M, de Paula Mota TR, Gauthier M (2005) Electrophysiological and behavioural characterization of gustatory responses to antennal 'bitter' taste in honeybees. Eur J Neurosci 22:3161–3170

de Brito Sanchez MG, Chen C, Li J, Liu F, Gauthier M, Giurfa M (2008) Behavioral studies on tarsal gustation in honeybees: sucrose responsiveness and sucrose-mediated olfactory conditioning. J Comp Physiol A 194:861–869

Denker M, Finke R, Schaupp F, Grün S, Menzel R (2010) Neural correlates of odor learning in the honeybee antennal lobe. Eur J Neurosci 31:119–133

Devaud JM, Blunk A, Podufall J, Giurfa M, Grünewald B (2007) Using local anaesthetics to block neuronal activity and map specific learning tasks to the mushroom bodies of an insect brain. Eur J Neurosci 26:3193–3206

Deisig N, Lachnit H, Hellstern F, Giurfa M (2001) Configural olfactory learning in honeybees: negative and positive patterning discrimination. Learn Mem 8:70–78

Deisig N, Lachnit H, Giurfa M (2002) The effect of similarity between elemental stimuli and compounds in olfactory patterning discriminations. Learn Mem 9:112–121

Deisig N, Lachnit H, Sandoz JC, Lober K, Giurfa M (2003) A modified version of the unique cue theory accounts for olfactory compound processing in honeybees. Learn Mem 10:199–208

Dupuy F, Sandoz JC, Giurfa M, Josens R (2006) Individual olfactory learning in *Camponotus* ants. Anim Behav 72:1081–1091

Erber J, Masuhr T, Menzel R (1980) Localization of short-term memory in the brain of the bee, *Apis mellifera*. Physiol Entomol 5:343–358

Faber T, Joerges J, Menzel R (1999) Associative learning modifies neural representations of odors in the insect brain. Nat Neurosci 2:74–78

Farina WM, Núñez JA (1991) Trophallaxis in the honeybee, Apis mellifera(L) as related to the profitability of food sources. Anim Behav 42:389–394

Friedrich A, Thomas U, Müller U (2004) Learning at different satiation levels reveals parallel functions for the cAMP-protein kinase A cascade in formation of long-term memory. J Neurosci 24:4460–4468

Frings H (1944) The loci of olfactory end-organs in the honey-bee, *Apis mellifera* Linn. J Exp Zool 97:123–134

Frings H, Frings M (1949) The loci of contact chemoreceptors in insects. A review with new evidence. Am Midl Nat 41:602–658

Frost EH, Shutler D, Hillier NK (2011) Effects of cold immobilization and recovery period on honeybee learning, memory, and responsiveness to sucrose. J Insect Physiol 57:1385–1390

Galizia CG, Joerges J, Küttner A, Faber T, Menzel R (1997) A semi-in-vivo preparation for optical recording of the insect brain. J Neurosci Methods 76:61–69

Gauthier M, Dacher M, Thany SH, Niggebrugge C, Deglise P, Kljucevic P, Armengaud C, Grunewald B (2006) Involvement of α-bungarotoxin-sensitive nicotinic receptors in long-term memory formation in the honeybee (*Apis mellifera*). Neurobiol Learn Mem 86:164–174

Gerber B, Ullrich J (1999) No evidence for olfactory blocking in honeybee classical conditioning. J Exp Biol 202:1839–1854

Giurfa M (2007) Behavioral and neural analysis of associative learning in the honeybee: a taste from the magic well. J Comp Physiol A 193:801–824

Giurfa M, Malun D (2004) Associative mechanosensory conditioning of the proboscis extension reflex in honeybees. Learn Mem 11:294–302

Giurfa M, Menzel R (1997) Insect visual perception: complex abilities of simple nervous systems. Curr Opin Neurobiol 7:505–513

Giurfa M, Lehrer M (2001) Honeybee vision and floral displays: from detection to close-up recognition. In: Chittka L, Thomson J (eds) Cognitive ecology of pollination. Cambridge University Press, Cambridge, pp 61–82

Giurfa M, Sandoz JC (2012) Invertebrate learning and memory: fifty years of olfactory conditioning of the proboscis extension response in honeybees. Learn Mem 19:54–66

Giurfa M, Eichmann B, Menzel R (1996) Symmetry perception in an insect. Nature 382:458–461

Giurfa M, Zhang S, Jennet A, Menzel R, Srinivasan MV (2001) The concepts of "sameness" and "difference" in an insect. Nature 410:930–933

Gronenberg W (2001) Subdivisions of hymenopteran mushroom body calyces by their afferent supply. J Comp Neurol 436:474–489

Grünbaum L, Müller U (1998) Induction of a specific olfactory memory leads to a long-lasting activation of protein kinase C in the antennal lobe of the honeybee. J Neurosci 18:4384–4392

Grünewald B (1999) Physiological properties and response modulations of mushroom body feedback neurons during olfactory learning in the honeybee, *Apis mellifera*. J Comp Physiol A 185:565–576

Guerrieri F, Schubert M, Sandoz JC, Giurfa M (2005a) Perceptual and neural olfactory similarity in honeybees. PLoS Biol 3:e60. doi:10.1371/journal.pbio.0030060

Guerrieri F, Lachnit H, Gerber B, Giurfa M (2005b) Olfactory blocking and odorant similarity in the honeybee. Learn Mem 12:86–95

Hadar R, Menzel R (2010) Memory formation in reversal learning of the honeybee. Front Behav Neurosci 4:186. doi:10.3389/fnbeh.2010.00186

Haehnel M, Menzel R (2010) Sensory representation and learning-related plasticity in mushroom body extrinsic feedback neurons of the protocerebral tract. Front Syst Neurosci 4:61. doi:10.3389/fnsys.2010.00161

Hammer M (1993) An identified neuron mediates the unconditioned stimulus in associative olfactory learning in honeybees. Nature 366:59–63

Hammer M, Menzel R (1995) Learning and memory in the honeybee. J Neurosci 15:1617–1630

Hammer M, Menzel R (1998) Multiple sites of associative odor learning as revealed by local brain microinjection of octopamine in honeybees. Learn Mem 5:146–156

Hammer TJ, Hata C, Nieh JC (2009) Thermal learning in the honeybee, *Apis mellifera*. J Exp Biol 212:3928–3934

Hori S, Takeuchi H, Arikawa K, Kinoshita M, Ichikawa N, Sasaki M, Kubo T (2006) Associative visual learning, color discrimination, and chromatic adaptation in the harnessed honeybee *Apis mellifera* L. J Comp Physiol A 192:691–700

Hori S, Takeuchi H, Kubo T (2007) Associative learning and discrimination of motion cues in the harnessed honeybee *Apis mellifera* L. J Comp Physiol A 193:825–833

Hosler JS, Smith BH (2000) Blocking and the detection of odor components in blends. J Exp Biol 203:2797–2806

Hourcade B, Muenz TS, Sandoz JC, Rößler W, Devaud JM (2010) Long-term memory leads to synaptic reorganization in the mushroom bodies: a memory trace in the insect brain? J Neurosci 30:6461–6465

Hussaini SA, Komischke B, Menzel R, Lachnit H (2007) Forward and backward second-order Pavlovian conditioning in honeybees. Learn Mem 14:678–683

Komischke B, Giurfa M, Lachnit H, Malun D (2002) Successive olfactory reversal learning in honeybees. Learn Mem 9:122–129

Kuwabara M (1957) Bildung des bedingten Reflexes von Pavlovs Typus bei der Honigbiene, *Apis mellifica*. J Fac Sci Hokkaido Univ Ser VI Zool 13:458–464

Lehrer M (1997) Honeybee's visual orientation at the feeding site. In: Leher M (ed) Orientation and communication in arthropods. Birkhäuser, Basel, pp 115–144

Lehmann M, Gystav D, Galizia CG (2011) The early bee catches the flower—circadian rhythmicity influences learning performance in honey bees, *Apis mellifera*. Behav Ecol Sociobiol 65:205–215

Locatelli F, Bundrock G, Müller U (2005) Focal and temporal release of glutamate in the mushroom bodies improves olfactory memory in *Apis mellifera*. J Neurosci 25:11614–11618

Lunney GH (1970) Using analysis of variance with a dichotomous dependent variable: an empirical study. J Educ Meas 7:263–269

Maleszka R, Helliwell P, Kucharski R (2000) Pharmacological interference with glutamate reuptake impairs long-term memory in the honeybee, *Apis mellifera*. Behav Brain Res 115:49–53

Matsumoto Y, Mizunami M (2000) Olfactory learning in the cricket *Gryllus bimaculatus*. J Exp Biol 203:2581–2588

Matsumoto Y, Mizunami M (2002) Temporal determinants of olfactory long-term retention in the cricket *Gryllus bimaculatus*. J Exp Biol 205:1429–1437

Mauelshagen J (1993) Neural correlates of olfactory learning in an identified neuron in the honeybee brain. J Neurophysiol 69:609–625

Menzel R (1990) Learning, memory, and "cognition" in honeybees. In: Kesner Rp Olten DS (ed) Neurobiology of comparative cognition. Erlbaum, Hillsdale, NJ, pp 237–292

Menzel R (1999) Memory dynamics in the honeybee. J Comp Physiol A 185:323–340

Menzel R (2001) Searching for the memory trace in a mini-brain, the honeybee. Learn Mem 8:53–62

Menzel R, Erber J, Masuhr T (1974) Learning and memory in the honeybee. In: Barton-Browne L (ed) Experimental analysis of insect behavior. Springer, Berlin, pp 195–217

Menzel R, Greggers U, Hammer M (1993) Functional organization of appetitive learning and memory in a generalist pollinator, the honeybee. In: Lewis AC (ed) Insect Learning. Chapman Hall, London, pp 79–125

Menzel R, Manz G, Menzel R, Greggers U (2001) Massed and spaced learning in honeybees: the role of CS, US, the intertrial interval, and the test interval. Learn Mem 8:198–208

Minnich DE (1921) An experimental study of the tarsal chemoreceptors of two nymphalid butterflies. J Exp Zool 33:173–203

Minnich DE (1926) The organs of taste on the proboscis of the blowfly *Phormia regina* Meigen. Anat Rec 34:126

Mizunami M, Yokohari F, Takahata M (2004) Further exploration into the adaptive design of the arthropod "microbrain": I. Sensory and memory-processing systems. Zool Sci 21:1141–1151

Mobbs PG (1982) The brain of the honeybee *Apis mellifera*. I. The connections and spatial organization of the mushroom bodies. Philos Trans R Soc B 298:309–354

Müller D, Gerber B, Hellstern F, Hammer M, Menzel R (2000) Sensory preconditioning in honeybees. J Exp Biol 203:1351–1364

Müller U (1996) Inhibition of nitric oxide synthase impairs a distinct form of long-term memory in the honeybee, *Apis mellifera*. Neuron 16:541–549

Müller U (2000) Prolonged activation of cAMP-dependent protein kinase during conditioning induces long-term memory in honeybees. Neuron 27:159–168

Müßig L, Richlitzsk A, Rößler R, Eisenhardt D, Menzel R, Leboulle G (2010) Acute disruption of the NMDA receptor subunit NR1 in the honeybee brain selectively impairs memory formation. J Neurosci 30:7817–7825

Okada R, Rybak J, Manz G, Menzel R (2007) Learning-related plasticity in PE1 and other mushroom body-extrinsic neurons in the honeybee brain. J Neurosci 27:11736–11747

Pamir E, Chakroborty NK, Stollhoff N, Gehring KB, Antemann V, Morgenstern L, Felsenberg J, Eisenhardt D, Menzel R, Nawrot MP (2011) Average group behavior does not represent individual behavior in classical conditioning of the honeybee. Learn Mem 18:733–741

Pavlov IP (1927) Lectures on conditioned reflexes. International publishers, New York

Perisse E, Raymond VD, Néant I, Matsumoto Y, Leclerc C, Moreau M, Sandoz JC (2009) Early calcium increase triggers the formation of olfactory long-term memory in honeybees. BMC Biol 7:30. doi:10.1186/1741-7007-7-30

Pham-Delègue MH, Bailez O, Blight MM, Masson C, Picard-Nizou AL, Wadhams LJ (1993) Behavioural discrimination of oilseed rape volatiles by the honey bee *Apis mellifera* L. Chem Senses 18:483–494

Ray S, Ferneyhough B (1997a) The effects of age on olfactory learning and memory in the honeybee *Apis mellifera*. Neuroreport 8:789–793

Ray S, Ferneyhough B (1997b) Seasonal variation of proboscis extension reflex conditioning in the honeybee, *Apis mellifera*. J Apic Res 36:108–110

Ray S, Ferneyhough B (1999) Behavioral development and olfactory learning in the honeybee (*Apis mellifera*). Dev Psychobiol 34:21–27

Sakura M, Okada R, Aonuma H (2012) Evidence for instantaneous e-vector detection in the honeybee using an associative learning paradigm. Proc R Soc B 279:535–542

Sandoz JC (2011) Behavioral and neurophysiological study of olfactory perception and learning in honeybees. Front Syst Neurosci 5:98. doi:10.3389/fnsys.2011.00098

Sandoz JC, Pham-Delègue MH (2004) Spontaneous recovery after extinction of the conditioned proboscis extension response in the honeybee. Learn Mem 11:586–597

Scheiner R, Page RE, Erber J (2001) Responsiveness to sucrose affects tactile and olfactory learning in preforaging honey bees of two genetic strains. Behav Brain Res 120:67–73

Seeley TD (1982) Adaptive significance of the age polyethism schedule in honeybee colonies. Behav Ecol Sociobiol 11:287–293

Seeley TD (1995) The wisdom of the hive-the social physiology of honey bee colonies. Harvard University Press, London

Si A, Hlliwell P, Maleszka R (2004) Effects of NMDA receptor antagonists on olfactory learning and memory in the honeybee (*Apis mellifera*). Pharmacol Biochem Behav 77:191–197

Smith BH (1998) Analysis of interaction in binary odorant mixtures. Physiol Behav 65:397–407

Smith BH, Cobey S (1994) The olfactory memory of the honeybee *Apis mellifera*. II. blocking between odorants in binary mixtures. J Exp Biol 195:91–108

Stach S, Benard J, Giurfa M (2004) Local-feature assembling in visual pattern recognition and generalization in honeybees. Nature 429:758–761

Stollhoff N, Menzel R, Eisenhardt D (2005) Spontaneous recovery from extinction depends on the reconsolidation of the acquisition memory in an appetitive learning paradigm in the honeybee (*A. mellifera*). J Neurosci 25:4485–4492

Stollhoff N, Menzel R, Eisenhardt D (2008) One retrieval trial induces reconsolidation in an appetitive learning paradigm in honeybees (*A. mellifera*). Neurobiol Learn Mem 89:419–425

Strube-Bloss MF, Nawrot MP, Menzel R (2011) Mushroom body output neurons encode odor-reward associations. J Neurosci 31:3129–3140

Szyszka P, Galkin A, Menzel R (2008) Associative and non-associative plasticity in kenyon cells of the honeybee mushroom body. Front Sys Neurosci 2:3. doi:10.3389/neuro.06.003.2008

Takeda K (1961) Classical conditioned response in the honey bee. J Insect Physiol 6:168–179

The Honeybee Genome Sequencing Consortium (2006) Insights into social insects from the honeybee *A. mellifera*. Nature 443:931–949

Tully T, Quinn WG (1985) Classical conditioning and retention in normal and mutant *Drosophila melanogaster*. J Comp Physiol A 157:263–277

Vergoz V, Roussel E, Sandoz JC, Giurfa M (2007) Aversive learning in honeybees revealed by the olfactory conditioning of the sting extension reflex. PLoS One 2:e288. doi:10.1371/journal.pone.0000288

von Frisch K (1914) Der Farbensinn und Formensinn der Biene. Zool Jahrb Physiol 37:1–238

von Frisch K (1967) The dance language and orientation of bees. Belknap, Cambridge

Wehner R (1981) Spatial vision in arthropods. In: Autrum HJ (ed) Handbook of sensory physiology Vic. Springer, Berlin, pp 287–616

Wright GA, Mustard JA, Simcock NK, Ross-Taylor AAR, McNicholas LD, Popescu A, Marion-Pol F (2010) Parallel reinforcement pathways for conditioned food aversions in the honeybee. Curr Biol 20:2234–2240

Part II
Electrophysiology

Chapter 3
Mining Spatio-Spectro-Temporal Cortical Dynamics: A Guideline for Offline and Online Electrocorticographic Analyses

Mining Cortical Dynamics from ECoG Data

Zenas C. Chao and Naotaka Fujii

Abstract Recent advances in the technology of electrocorticography (ECoG) allow accessing neural activity from most of the cortex, which poses the challenge of extracting relevant information from an overwhelming amount of data. In this chapter, we will present useful routines for identifying statistically significant features in high-dimensional ECoG signals (offline analysis) and for establishing decoding models that can translate ECoG signals to specific behavioral measures in real time (online analysis). We will use our data, which are freely available online, in a step-by-step demonstration and will highlight useful MATLAB toolboxes for trouble-free implementation.

Keywords Connectivity • Cortical dynamics • Data mining • Electrocorticography (ECoG) • MATLAB • Offline analysis • Online analysis • Time–frequency representation

3.1 Introduction

Electrocorticography (ECoG) records brain activity using grids or stripes of electrodes implanted on the surface of the cortex. As most large cortical neurons are oriented perpendicular to the cortical surface, correlated activity within a cortical column should sum similarly. Consequently, ECoG is most favorable, among the brain activity recording techniques available, for measuring this correlated activity, especially across a wide area.

Z.C. Chao • N. Fujii (✉)
Laboratory for Adaptive Intelligence, RIKEN Brain
Science Institute, Saitama, Japan
e-mail: na@brain.riken.jp

In contrast to single-unit activity (SUA) recordings, ECoG measures population activity, which offers a better prospect for long-term recording stability (Chao et al. 2010). Furthermore, as ECoG recording does not penetrate the cortex, signal-prohibitive encapsulation, an obstacle in chronic SUA, multiunit activity (MUA), and local field potential (LFP) recordings, is less likely to occur during long-term implantation (Vetter et al. 2004; Szarowski et al. 2003; Bjornsson et al. 2006). Compared with electroencephalography (EEG), ECoG offers higher spatial resolution, broader bandwidth, and higher amplitude and is less sensitive to artifacts such as electromyogram (EMG) (Freeman et al. 2003; Schwartz et al. 2006). Compared with functional magnetic resonance imaging (fMRI) and near-infrared spectroscopy (NIRS), which are based on the blood oxygenation level, ECoG offers direct measures of neural activity with significantly higher temporal resolution. In conclusion, ECoG provides a great balance between signal fidelity, temporal and spatial resolutions, long-term durability and stability, and capability to cover multiple brain regions. Therefore, ECoG may be the optimal choice for mining spatio-spectro-temporal cortical dynamics.

Successes in ECoG research have accumulated greatly during the past decade in the areas of neuroscience and neuroengineering and especially in the field of brain–machine interfaces (BMIs) [see reviews in (Donoghue 2002; Mussa-Ivaldi and Miller 2003; Nicolelis 2003; Lebedev and Nicolelis 2006; Patil and Turner 2008)]. One recent development of ECoG recording technology enables 288-channel recordings with submillimeter and submillisecond resolution (Viventi et al. 2010). Our laboratory also developed a 256-channel ECoG system that can cover most of the cortex, including structures in the medial wall (Nagasaka et al. 2011). These advances pose a challenge to data analysis regarding how to extract relevant information from terabytes of data with satisfactory thoroughness and efficiency.

Depending on the goal of specific studies, ECoG analysis can be generally classified into two main categories: offline and online analyses. Offline analysis aims to identify statistically relevant features in ECoG signals underlying the neural processes of interest (Sect. 3.2). Conversely, online analysis focuses on establishing a decoding model that can translate ECoG signals to specific sensory inputs, motor outputs, or cognitive processes in real time (Sect. 3.3). Online analysis is usually, but not exclusively, used in BMI applications, where real-time interpretation of brain activity is used for either controlling external devices or estimating cognitive states. Both offline and online analyses can provide insights into how the brain encodes information, and both require the mining of high-dimensional data to extract relevant characteristics in the signals.

Offline and online analyses are not unique to ECoG signals. Many tools used in ECoG analysis originate from EEG analysis and are shared with analyses of fMRI and other electrophysiological technologies, such as magnetoencephalography (MEG). Here, we will not focus on detailed theoretical backgrounds. Instead, our goals are to provide the following:

- Routines for mining multichannel ECoG signals, particularly in the frequency domain, as frequency bands in ECoG signals have distinctive functional interpretations

- Comparisons of commonly used techniques and highlights of their advantages and shortcomings
- Demonstrations using data from our laboratory that are freely available (neurotycho.org) (Nagasaka et al. 2011)
- Useful software, MATLAB (The MathWorks, Inc.) toolboxes in particular, for easier implementation

3.2 Offline Analysis

The goal of offline analysis is to extract characteristics in ECoG signals that are statistically relevant to the neural processes of interest. In this section, we will go through a routine step-by-step approach for identifying useful features from multi-trial, multichannel ECoG data in a typical block-design experiment (Fig. 3.1).

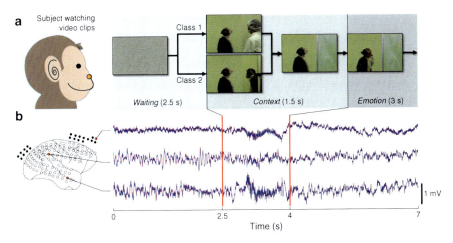

Fig. 3.1 Block-design experiment for social-context-dependent fear recognition: (**a**) The subject, a Japanese monkey (*left*), was seated in front of a TV screen showing a series of 7 s movie clips (*right*). Two types of clips were used, each consisting of three parts. *Waiting* period (2.5 s): only white noise was shown. *Context* period (1.5 s): one of two types of social conditions was shown—a monkey is threatened by a human (Class 1) or by another monkey (Class 2). At the end of this period, a curtain was closed and only the monkey being threatened could be seen by the subject. *Emotion* period (3 s): the monkey being threatened in the *Context* period exhibited an expression of fear. The two types of clips were played in random order and each was played a total of 100 times. The goal was to identify the differences in ECoG signals recorded in the subject watching the same fear expression but under different social contexts, which should represent the neural correlates of social-context-dependent fear recognition, i.e., how the subject perceives the fear expression shown by another monkey based on different social contexts. (**b**) A 128-channel ECoG array was implanted into the subject (*left*), which covered most of the right hemisphere, including the medial wall in the frontal and the visual cortices (*black dots*). Voltage traces, sampled at 1 kHz, of a randomly picked trial at three electrodes (*red dots*) are shown (*right*). The onsets of the *Context* and *Emotion* periods are indicated as two *red vertical lines*

In multichannel recordings, the dimension of the information is commonly not as high as the number of channels. Hence, independent component analysis (ICA) is usually introduced to identify underlying signal sources (Sect. 3.2.1). As a result, ICA can lead to artifacts extraction and data dimension reduction, which are beneficial for the subsequent calculation of activity features. Two types of activity features are widely used to quantify event-related spatio-spectro-temporal brain dynamics: time–frequency representations (TFRs, Sect. 3.2.2) and connectivity measures (Sect. 3.2.3). The two features offer different information outlines that complement each other. The last step in offline analysis is the identification of statistically significant features underlying the target neural processes (Sect. 3.2.4).

3.2.1 Independent Component Analysis

ICA is a technique that is used to separate multichannel signals into their constituent underlying sources, such that each source contributes as much distinct/independent information to the data as possible. Adding this step to ECoG analysis may provide lower-dimensional, nonartifactual, and statistically independent source signals. Consequently, ICA can reduce the computational load and prevent colinearity in the data, which are essential for calculating more complex activity features, such as connectivity measures (Sect. 3.2.3).

Preprocessing Before ICA

Data preprocessing is the most overlooked step and could diminish the effectiveness of ICA or even lead to spurious results if not performed carefully. The most crucial step in the preprocessing step before performing ICA is to reject artifacts in the raw signals. For multitrial data, such as the data from the fear recognition experiment, a thorough strategy for artifact rejection can be found in EEGLAB wiki (Delorme and Makeig 2004). The basic concept is to reject trials or channels that appear to contain artifacts using visual inspection, statistical thresholding, or a combination of both. To analyze the fear recognition data, we rejected channels and trials with abnormal spectra, which has been suggested as the most effective method (Delorme et al. 2001). An alternative strategy is to select nonartifactual independent components (ICs) after performing ICA on the raw data.

Model Order and Component Selections

When the number of channels exceeds the number of real sources, the model order, i.e., the number of ICs to estimate, can have a significant impact on the quality and accuracy of ICA. If the model order is greater than the actual number of sources, overfitting of the ICA model could occur and lead to distortions (Hyvärinen and Oja 2000). The most commonly used criterion for selecting the model order for ICA is based on principal component analysis (PCA) of the data covariance matrix, in which the model order is set as the number of eigenvalues that account for a certain proportion of the total observed variance. For the fear recognition data, more than 75 % of the total variance can be explained by 20 ICs.

It is also useful, sometimes essential, to select a subset of ICs that are related to brain activity. This can be executed manually (Onton and Makeig 2009) or by using automatic algorithms when characteristics of artifacts are known in prior, such as ADJUST (Mognon et al. 2011), which can be implemented using the EEGLAB library (Delorme et al. 2011).

ICA Algorithms

ICA can be classified into two main categories according to the algorithmic approach: one based on higher-order statistics (HOS) and another based on the time structure [see review in (James and Hesse 2005)]. HOS-based ICA seeks statistical independence in sources, not through uncorrelatedness, as in PCA, but through HOS, such as kurtosis and differential entropy. In contrast to HOS-based ICA, in which temporal ordering of the signals is irrelevant, time-structure-based ICA seeks independence in sources through temporal or spectro-temporal uncorrelatedness. In comparison, HOS-based ICA seems less intuitive, as it discards the temporal information of the data, which is apparently relevant in ECoG signals. Moreover, time-structure-based ICA can be adapted to deal with nonstationary signals (James and Hesse 2004). However, HOS-based ICA is more efficient computationally and, thus, more practical when the number of channels and/or signal length is large.

The three widely used HOS-based ICA algorithms are (1) FastICA, which aims to maximize the magnitude of kurtosis to render the sources as independent as possible (Hyvärinen and Oja 1997); (2) Infomax, a neural network gradient-based algorithm that attempts to measure independence using differential entropy (Bell and Sejnowski 1995); and (3) JADE, a tensorial method that uses higher-order cumulant tensors (Cardoso and Souloumiac 1993). The three algorithms yield similar results in the fear recognition data (not shown) and in fMRI data (Zibulevsky and Pearlmutter 2001); however, JADE possesses a limitation on high-dimensional data, for numerical reasons. One straightforward approach that can be used for time-structure-based ICA is based on spatio-temporal decorrelation, which maximizes the independence between sources in the time domain (Ziehe and Müller 1998; Belouchrani et al. 1997). This method is flexible and more effective for the extraction of neurophysiologically meaningful components in short segment data, where HOS-based ICA failed (James and Hesse 2004), and can be implemented using ICALAB toolboxes (Cichocki et al. 2002). Another attractive approach that is particularly suitable for ECoG signals is based on spatio-spectro-temporal decorrelation, which maximizes the independence between sources in both time and frequency domains. However, this method can be prohibitive, even for small data sets, because of its intensive memory usage. Some modifications have been attempted to improve its efficiency; one example is the TFBSS toolbox (Févotte and Doncarli 2004).

Reliability of ICA Estimations

One major issue in the application of ICA is that the reliability of the estimated ICs is not known. In practice, most algorithms may give different results when run multiple times, as only the local minimum of the objective, the independence measure to be optimized, is found in each run. One solution is to run the ICA algorithm many times using different initial values and different bootstrapped data sets (Meinecke et al. 2002), which can be implemented using the ICASSO package (Himberg et al. 2004). If an IC is reliable, the results from multiple runs should yield a cluster that is close to the ideal component. Based on this concept, we can further evaluate the quality of each component. The results of the implementation of ICASSO with the FastICA algorithm on our fear recognition data are shown in Fig. 3.2.

3.2.2 Time–Frequency Representations

When analyzing neurological signals, TFRs are useful for investigating spectral contents in addition to studying changes in its time domain features. Two TFRs are widely used in the analysis of ECoG signals: the spectrogram [the squared magnitude of the short-time Fourier transform (STFT)] and the scalogram [the squared magnitude of the continuous wavelet transform (CWT)].

For the spectrogram, STFT modulates the signal with a window function, commonly a Hann window, before performing the Fourier transform to obtain the frequency content in the region of the window. This method is straightforward but has its own drawbacks of leakage effects and limitation of uniform resolution. Spectral leakage, which causes false frequency components, could be reduced by implementing the multitaper method, which reduces estimation bias by obtaining multiple independent estimates from the same sample (Thomson 1982); this can be implemented using the Chronux library (Bokil et al. 2010). Furthermore, the constant-length windows used in STFT result in a uniform partition in the spectrogram, which limits the analysis to a single resolution for the complete signal. This can be problematic, as most of the signals of practical interest have high-frequency components for short durations and low-frequency components for long durations. In this aspect, multiresolution TFR, such as the scalogram, is more desirable.

For the scalogram, CWT uses short windows at high frequencies and long windows at low frequencies and is more suitable for the analysis of nonstationary signals than is the STFT-based spectrogram (Huang et al. 1998; Mallat 1989). A brief overview of a wavelet-based time-scale analysis of biological signals is given in (Thakor and Sherman 1995). Scalogram calculation can be implemented using the Time–Frequency Toolbox developed by CNRS (Auger et al. 1999).

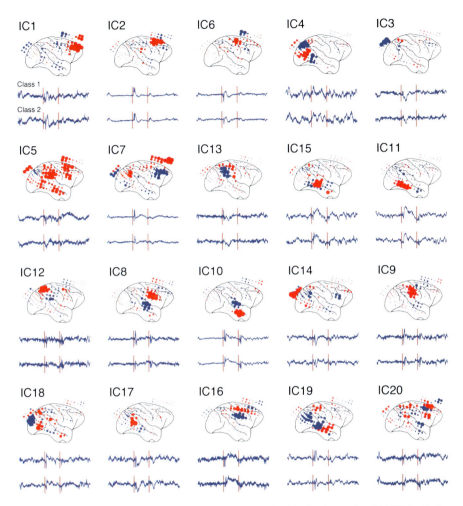

Fig. 3.2 ICA on the fear recognition data: ICs were acquired by implementing ICASSO with the FastICA algorithm, with 20 ICs and ~75 % of variance explained. The spatial distribution of the unmixing matrix of each component, which indicates how signals from different electrodes contribute to the component, is shown on the top of each panel. The size of each dot represents the normalized absolute contribution of each electrode to the component, and the color of each *dot* represents the corresponding sign (*red*, positive; *blue*, negative). The average IC across trials, or the event-related potential (ERP) of each component, for each class is also shown. The two *red vertical lines* represent the onset of events, as in Fig. 3.1. Twenty ICs (IC1–IC20) are shown in the order of the quality index representing the estimation reliability (Himberg et al. 2004)

3.2.3 Connectivity Measures

The primary goal of connectivity measures is to identify relationships between activities in different brain regions during information processing. Connectivity measures can reveal the structural or anatomical connectivity of the brain and provide insights into how different brain regions integrate information, which is lacking in TFR analysis (Sect. 3.2.2). Connectivity measures can be classified into two categories: (1) *functional* connectivity, which indicates symmetrical correlations in activity between brain regions, and (2) *effective* connectivity, which represents asymmetric causal dependencies in activity between brain regions (Bullmore and Sporns 2009).

In this section, we will outline some commonly used spectral connectivity measures for multitrial multichannel data, which can be derived from the coefficients of the vector autoregressive (VAR) model.[1] The accurate estimation of VAR coefficients requires that each time series be covariance stationary, i.e., its mean and variance remain unchanged over time. However, ECoG signals are usually highly nonstationary, exhibiting dramatic and transient fluctuations (see examples in Fig. 3.1b). Several methods have been proposed to improve stationarity [more detail in (Delorme et al. 2011)]. In this chapter, we will demonstrate connectivity measures by implementing a sliding-window method. The concept is to segment the signals into sufficiently small windows, then measure connectivity within each window, where the signal is *locally* stationary.

A detailed tutorial of VAR-based connectivity measures can be found in the Source Information Flow Toolbox (SIFT) handbook (http://sccn.ucsd.edu/wiki/SIFT). Most routines in this section were implemented using SIFT (Delorme et al. 2011) together with other libraries, such as EEGLAB (Delorme and Makeig 2004), Granger causal connectivity analysis (GCCA) (Seth 2010), and Brain-System for Multivariate AutoRegressive Timeseries (BSMART) (Cui et al. 2008).

Preprocessing Before Connectivity Measures

For multitrial data, the standard preprocessing steps for achieving local stationarity are (1) detrending, (2) temporal normalization, and (3) ensemble normalization (Ding et al. 2000). Detrending, which is the subtraction of the best-fitting line from each time series, removes the linear drift in the data. Temporal normalization, which is the subtraction of the mean of each time series and division by the standard deviation, ensures that all variables have equal weights across the trial. These processes should be performed on each trial for each channel. Ensemble normalization, which is the pointwise subtraction of the ensemble mean and division by the ensemble standard deviation, targets rich task-relevant information that cannot be inferred from the event-related potential (ERP) (Ding et al. 2000; Bressler and Seth 2011).

---

[1] A parametric model used to capture the linear interdependencies among multiple time series. A VAR model describes a set of variables over the sample period as a linear function of their past evolution, i.e., the variables of sample t as a linear combination of the variables of samples $[t-1, t-2,...,t-p]$, where p is referred to as the *model order*. The VAR coefficients, after being transformed into the frequency domain via Z transformation, can be used to compute spectral connectivity measures.

Window Length and Model Order Selections

The important considerations for determining window length for segmentation are the balance between (1) acquiring a sufficient number of data points and (2) maintaining local stationarity. The window length should be small enough to allow treating the data as stationary yet large enough to allow estimation of VAR coefficients. The general rule is that the number of parameters should be <10 % of the data samples, i.e., to fit a VAR model with model order p on data of k dimensions (k ICs selected from ICA), the following relation needs to be satisfied: $w \geq 10 \times (k^2 \times p/n)$, where w and n represent the window length and the number of trials, respectively.

Model order, which is related to the length of the signal in the past that is relevant to the current observation, is another key parameter that needs to be determined. If the chosen model is too low, the VAR model cannot capture dynamic relations in the data and the frequency resolution will be impaired. The most popular approach for model order selection is based on the Akaike information criterion (AIC) (Akaike 1974) and/or the Bayesian information criterion (BIC) (Schwarz 1978). A detailed comparison of different information criteria can be found in (Lütkepohl 2005).

Spectral Functional Connectivity Measures

Functional connectivity, which is defined as the temporal correlation between spatially remote neurophysiological events (Friston et al. 1993), is one way to characterize interaction or functional integration between brain regions. Coherency, which is the spectral analogue of the cross correlation, and its derivatives, such as coherence and partial coherence, are common spectral VAR-based functional connectivity measures (Brillinger 2001). Their incapability to access causality or directionality between or among signals is one of the main drawbacks of functional connectivity measures; this can be overcome by using effective connectivity measures.

Spectral Effective Connectivity Measures

Most of the spectral effective connectivity measures are based on the concept of Granger causality (GC): if one can predict signal X better by incorporating past information from signal Y than by using only information from its own past, then signal Y is causal for signal X (Wiener 1956; Granger 1969). Brief overviews of effective connectivity measures for multichannel data can be found in (Delorme et al. 2011; Blinowska 2011).

In contrast to functional connectivity measures, effective connectivity measures capture asymmetric causal dependencies between signals, which can reveal directional information flow between brain regions. Effective connectivity measures can be classified into two types, according to the number of signals included in the VAR estimation: bivariate or multivariate. Bivariate measures estimate VAR coefficients for each signal pair, one at a time; thus, they are likely to be immune from a shortage of computational resources. However, they cannot distinguish between *direct* and *indirect* interactions, i.e., whether the information flow between two signals is mediated by a third signal (Kus et al. 2004). This can be overcome by using multivariate measures, in which VAR coefficients from multiple (>2) signals are estimated simultaneously. However, significantly more computational resources will be required when the number of signals increases. Thus, dimension reduction techniques, such as ICA, are essential for data mining with a large number of channels.

Granger–Geweke causality (GGC), or frequency-domain GC, is a commonly used bivariate spectral effective connectivity measure (Geweke 1982; Bressler et al. 2007). Thorough discussions of GGC in neurophysiological data can be found in (Ding et al. 2006; Scott and Kalaska 1997). For multivariate effective connectivity, we will focus primarily on three measures: partial directed coherence (PDC) (Baccala and Sameshima 2001), direct directed transfer function (dDTF) (Korzeniewska et al. 2003), and renormalized PDC (rPDC) (Schelter et al. 2009). The commonly used PDC and dDTF are multivariate extensions of the GC concept and can be interpreted as frequency-domain conditional GC, in which all signals are taken into account to disentangle direct and indirect influences among signal pairs. rPDC is a scale-free measure that is independent of the units and variance of the signals and can be used to assess the strength of direct influences (Schelter et al. 2009).

3.2.4 Statistical Significance

The probability distributions of the activity features illustrated earlier are often not known. Nonparametric statistical methods are usually employed to test their statistical significance, to allow further inferences. This is particularly important and is usually standard for computing the significance level of connectivity measures. Here, we will introduce some common numerical methods for computing statistical significance and discuss their reliability, particularly on connectivity measures.

Numerical Approaches for Statistical Significance

Two methods are commonly used to obtain confidence intervals and statistical significance thresholds for activity features: bootstrap resampling (Efron and Tibshirani 1993) and the leave-one-out method (LOOM) (Schlögl and Supp 2006). Bootstrap resampling is a numerical method that is used to obtain a significant estimate by generating surrogate data through randomly resampling the data with replacement. This sampling with replacement is repeated many times (100 or 1,000) to approximate the true distribution. LOOM, which is based on jackknifing (Quenouille 1949),

3 Mining Spatio-Spectro-Temporal Cortical Dynamics... 49

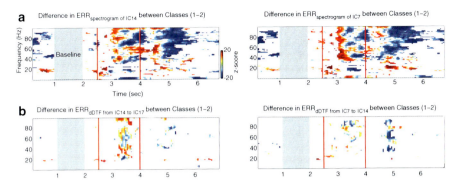

Fig. 3.3 Statistical significance in spectrograms and dDTFs: (**a**) Significant differences in event-related responses (ERRs) in spectrograms between Classes 1 and 2. Examples from IC14 (*left*) and IC7 (*right*) are shown. ERRs were obtained by comparing each value to a baseline (*gray*) for each frequency bin (a total of 65 bins from 0 to 100 Hz). Significant ERRs from two classes were then compared with each other. Significant differences in ERRs are shown as a *z*-score, e.g., areas with *reddish colors* indicate that ERRs in Class 1 are significantly greater than those in Class 2. Variables that were either not significant ERRs or not significantly different between two classes are shown in *white*. The Bonferroni correction was used to test ERRs and differences in ERRs. The two *red vertical lines* represent the onset of events, as in Fig. 3.1. (**b**) Significant differences in ERRs in dDTFs between Classes 1 and 2. Examples of dDTF from IC14 to IC7 (*left*) and dDTF from IC7 to IC14 (*right*) are shown. The methods and representations are as in (**a**)

measures activity features from multitrial data, leaving out a different trial each time. This method is straightforward but usually requires a large number of trials for practical applications.

The examination of the empirical distribution of each value in activity features after bootstrapping or LOOM allows us to perform various statistical analyses: (1) to test for significantly nonzero features; (2) to test for significant event-related responses, i.e., features that are significantly different from those in a baseline; and (3) to test for significant differences in features between experimental conditions. As those tests are individually performed on each value of an activity feature, it is important to correct for multiple comparisons, to acquire statistical significance on the complete feature. Two common correction methods are (1) the Bonferroni correction, which lowers the significance threshold (e.g., 0.05) $1/N$ times, where N represents the total number of variables in the activity feature, and (2) false discovery rate (FDR) (Benjamini and Hochberg 1995), which controls the expected proportion of incorrectly rejected null hypotheses (type I errors). Typically, FDR is less strict than the Bonferroni correction while still maintaining control for multiple comparisons. For our fear recognition data, we tested for significant event-related responses that are also significantly different among classes. Examples from two ICs (see Fig. 3.2) are shown in Fig. 3.3.

Reliability of Connectivity Measures

One important question when analyzing connectivity is which connectivity measures can capture "true" underlying interactions between or among signals. Although all measures have been initially tested in simulations and applied to neurophysiological data, the answer to this question remains unclear because of the lack of a thorough systematic comparison. A recent development tested different multivariate effective connectivity measures in combination with different numerical significance computational approaches on simulated data with various noise levels, data lengths, model orders, and coupling strengths (Florin et al. 2011). The results suggest that the squared magnitude of PDC combined with LOOM is the most reliable and robust choice. However, it is important to keep in mind that LOOM is problematic for data with a small number of trials. Nonetheless, choices of features and parameters are crucial, and great caution should be taken in practice.

3.3 Online Analysis

The goal of online analysis is to map activity features (observable variables) to the underlying neural processes of interest (predicted variables) in real time, which is designed primarily for BMI applications. Similar to offline analysis, one crucial step of online analysis is *feature selection*, i.e., to identify the relevant subset of activity features. Another important step is the design of a *decoder* (regressor or classifier) that can efficiently translate the selected features to continuous variables (e.g., movement trajectories) or discrete variables (e.g., cognitive states). Based on different strategies of feature selection and decoder design, the following approaches are usually implemented for high-dimensional data: the filter approach (Sect. 3.3.1), the wrapper approach (Sect. 3.2), and the integration approach (Sect. 3.3.3).

3.3.1 Filter Approach

In the filter approach, activity features are selected independent of decoding, and the selected features are then used in the design of a decoder. The features can be selected according to their relevance, such as correlation between each feature and the predicted variables (as in an example of ECoG-based hand-trajectory decoding Schalk et al. 2007), or by their statistical significance, as in the offline analysis described in Sect. 3.2. Typically, the filter approach is applied to univariate measures, such as TFRs; therefore, it does not consider the relationships between features while selecting them. As ECoG data are inherently multivariate, with strong spatial correlation between activities from different brain areas, it is also useful to select features based on connectivity measures. The fact that this method does not consider the performance of the decoder during feature selection is a drawback of the filter approach.

3.3.2 Wrapper Approach

The wrapper approach, in contrast to the filter approach, uses methods in which activity features are selected to maximize the performance of a decoder. Typically, features are included or excluded recursively, until the maximal performance of a predetermined decoder is reached, as in an example of ECoG-based motor imagery classification (Ince et al. 2009). The observations that the thresholds used to select features are usually arbitrary and the fact that different data sets may require different settings of thresholds are weaknesses of this approach (De Martino et al. 2008). Furthermore, the decoder needs repeated training in selecting features, which could be very time consuming.

3.3.3 Integration Approach

The integration approach simultaneously addresses the problem of feature selection and decoder design. In this strategy, feature selection is included as part of decoder design, ensuring efficient use of data and faster computation time, as the decoder does not need to be trained repeatedly during feature selection. This is particularly advantageous for a large number of features and limited training data.

Here, we will outline three multivariate approaches that do not require user-specified thresholds for feature selection and have demonstrated potential for decoding high-dimensional ECoG data. The first one is partial least squares regression (PLS-R), which aims to extract the latent structures in both observed and predicted variables, so that their relation is maximized in terms of the covariance structure. PLS-R can be implemented using the standard MATLAB Statistics toolbox. By using PLS-R on scalograms (see Sect. 3.2.2) of ECoG signals, we have successfully decoded high-dimensional continuous arm movements with high accuracy, which was comparable to SUA-based decoding (Chao et al. 2010) ([data are freely available at neurotycho.org Nagasaka et al. 2011). The second approach is sparse logistic regression (SL-R), which imposes sparsity constraints to obtain a small subset of nonzero model coefficients. This method has been used to classify whole-brain fMRI data (Ryali et al. 2010; Yamashita et al. 2008) and to reconstruct auditory stimuli from our 128-channel ECoG system (data not shown). SL-R can be implemented using the Sparse Logistic Regression (SLR) toolbox (Yamashita et al. 2008). The last, but not the least, potential approach is variational Bayesian least squares regression (VBLS-R), which is computationally efficient and suitable for large amounts of very high-dimensional data. VBLS-R was used to predict EMG activity and end-effector velocity from motor cortical activity (Ting et al. 2008). VBLS-R can be implemented using MATLAB codes downloadable from http://www-clmc.usc.edu/Resources/Software.

3.4 Conclusion

The brain encodes information in different modalities and in their integration. At the basic scientific level, ECoG is a balanced option to sample detailed brain activity and to extract this information. At the technological level, ECoG is capable of linking different recording techniques across the whole spectrum, with its correlation with other techniques being established, from SUA (Miller 2010) to EEG (Zhang et al. 2006) and fMRI (Conner et al. 2011). At the clinical level, ECoG provides supreme decoding performance and long-term stability for BMI applications, without damaging the brain.

Recent advances in ECoG technology have enabled the direct and simultaneous access to neural activity from most of the cortex, which poses the challenge of identifying relevant information from an overwhelming amount of data. The ideal approach to this challenge is to model the brain as a whole and to fit this model to a wide range of behaviors. Therefore, we can establish a general model that describes how the brain integrates information dynamically to represent not only one specific process but also the interplay among different processes. To achieve this ultimate goal, we will need to not only develop and evaluate different theoretical frameworks but also collect and share a variety of experimental data.

Acknowledgments We thank Yasuo Nagasaka, who helped design and collect the data described in this chapter, and Naomi Hasegawa and Tomonori Notoya, for their technical assistance.

References

Akaike H (1974) A new look at the statistical model identification. IEEE Trans Automat Cont 19(6):716–723
Auger F et al (1999) Time-frequency toolbox. http://tftb.nongnu.org
Baccala LA, Sameshima K (2001) Partial directed coherence: a new concept in neural structure determination. Biol Cybern 84(6):463–474
Bell AJ, Sejnowski TJ (1995) An information-maximization approach to blind separation and blind deconvolution. Neural Comput 7(6):1129–1159
Belouchrani A et al (1997) A blind source separation technique using second-order statistics. IEEE Trans Signal Process 45(2):434–444
Benjamini Y, Hochberg Y (1995) Controlling the false discovery rate: a practical and powerful approach to multiple testing. J R Stat Soc Series B (Methodol) 57(1):289–300
Bjornsson C et al (2006) Effects of insertion conditions on tissue strain and vascular damage during neuroprosthetic device insertion. J Neural Eng 3:196–207
Blinowska KJ (2011) Review of the methods of determination of directed connectivity from multichannel data. Med Biol Eng Comput 49(5):521–529
Bokil H et al (2010) Chronux: a platform for analyzing neural signals. J Neurosci Methods 192(1):146–151
Bressler SL, Seth AK (2011) Wiener–Granger causality: a well established methodology. Neuroimage 58(2):323–329
Bressler SL et al (2007) Cortical functional network organization from autoregressive modeling of local field potential oscillations. Stat Med 26(21):3875–3885

Brillinger DR (2001) Time series: data analysis and theory. SIAM, San Francisco
Bullmore E, Sporns O (2009) Complex brain networks: graph theoretical analysis of structural and functional systems. Nat Rev Neurosci 10(3):186–198
Cardoso JF, Souloumiac A (1993) Blind beamforming for non-Gaussian signals. IEE Proc Radar Signal Process 140(6):362–370
Chao ZC, Nagasaka Y, Fujii N (2010) Long-term asynchronous decoding of arm motion using electrocorticographic signals in monkeys. Front Neuroeng 3:3. doi:10.3389/fneng.2010.00003
Cichocki A et al (2002) ICALAB toolboxes. http://www.bsp.brain.riken.jp/ICALAB
Conner CR et al (2011) Variability of the relationship between electrophysiology and BOLD-fMRI across cortical regions in humans. J Neurosci 31(36):12855–12865
Cui J et al (2008) BSMART: a Matlab/C toolbox for analysis of multichannel neural time series. Neural Netw 21(8):1094–1104
De Martino F et al (2008) Combining multivariate voxel selection and support vector machines for mapping and classification of fMRI spatial patterns. Neuroimage 43(1):44–58
Delorme A, Makeig S (2004) EEGLAB: an open source toolbox for analysis of single-trial EEG dynamics including independent component analysis. J Neurosci Methods 134(1):9–21
Delorme A, Makeig S, Sejnowski T (2001) Automatic artifact rejection for EEG data using high-order statistics and independent component analysis. In: Proceedings of the third international ICA conference, San Diego, 2001
Delorme A et al (2011) EEGLAB, SIFT, NFT, BCILAB, and ERICA: new tools for advanced EEG processing. Comput Intell Neurosci 2011:130714
Ding M et al (2000) Short-window spectral analysis of cortical event-related potentials by adaptive multivariate autoregressive modeling: data preprocessing, model validation, and variability assessment. Biol Cybern 83(1):35–45
Ding M, Chen Y, Bressler SL (2006) Granger causality: basic theory and application to neuroscience. In: Schelter B, Winterhalder M, Timmer J (eds) Handbook of time series analysis: recent theoretical developments and applications. Wiley, Weinheim
Donoghue JP (2002) Connecting cortex to machines: recent advances in brain interfaces. Nat Neurosci 5:1085–1088
Efron B, Tibshirani RJ (1993) An introduction to the bootstrap. Chapman and Hall, New York
Févotte C, Doncarli C (2004) Two contributions to blind source separation using time-frequency distributions. IEEE Signal Process Lett 11(3):386–389
Florin E et al (2011) Reliability of multivariate causality measures for neural data. J Neurosci Methods 198(2):344–358
Freeman WJ et al (2003) Spatial spectra of scalp EEG and EMG from awake humans. Clin Neurophysiol 114(6):1053–1068
Friston K et al (1993) Functional connectivity: the principal-component analysis of large (PET) data sets. J Cereb Blood Flow Metab 13(1):5–14
Geweke J (1982) Measurement of linear dependence and feedback between multiple time series. J Am Stat Assoc 77(378):304–313
Granger CWJ (1969) Investigating causal relations by econometric models and cross-spectral methods. Econometrica 37(3):424–438
Himberg J, Hyvarinen A, Esposito F (2004) Validating the independent components of neuroimaging time series via clustering and visualization. Neuroimage 22(3):1214–1222
Huang NE et al (1998) The empirical mode decomposition and the Hilbert spectrum for nonlinear and non-stationary time series analysis. Proc R Soc Lond A Math Phys Eng Sci 454(1971):903–995
Hyvärinen A, Oja E (1997) A fast fixed-point algorithm for independent component analysis. Neural Comput 9(7):1483–1492
Hyvärinen A, Oja E (2000) Independent component analysis: algorithms and applications. Neural Netw 13(4–5):411–430
Ince NF, Goksu F, Tewfik AH (2009) ECoG based brain computer interface with subset selection. In: Fred A, Filipe J, Gamboa H (eds) Biomedical engineering systems and technologies. Springer, Berlin, pp 357–374

James CJ, Hesse CW (2004) A comparison of time structure and statistically based BSS methods in the context of long-term epileptiform EEG recordings. In: Puntonet CG, Prieto A (eds) Independent component analysis and blind signal separation (Lecture notes in computer science). Springer, Berlin, pp 1025–1032

James CJ, Hesse CW (2005) Independent component analysis for biomedical signals. Physiol Meas 26(1):R15–R39

Korzeniewska A et al (2003) Determination of information flow direction among brain structures by a modified directed transfer function (dDTF) method. J Neurosci Methods 125(1–2):195–207

Kus R, Kaminski M, Blinowska KJ (2004) Determination of EEG activity propagation: pair-wise versus multichannel estimate. IEEE Trans Biomed Eng 51(9):1501–1510

Lebedev MA, Nicolelis MAL (2006) Brain-machine interfaces: past, present and future. Trends Neurosci 29(9):536–546

Lütkepohl H (2005) New introduction to multiple time series analysis. Springer, Berlin

Mallat SG (1989) A theory for multiresolution signal decomposition: The wavelet representation. IEEE Trans Pattern Anal Mach Intell 11(7):674–693

Meinecke F et al (2002) A resampling approach to estimate the stability of one-dimensional or multidimensional independent components. IEEE Trans Biomed Eng 49(12):1514–1525

Miller KJ (2010) Broadband spectral change: evidence for a macroscale correlate of population firing rate? J Neurosci 30(19):6477–6479

Mognon A et al (2011) ADJUST: an automatic EEG artifact detector based on the joint use of spatial and temporal features. Psychophysiology 48(2):229–240

Mussa-Ivaldi FA, Miller LE (2003) Brain-machine interfaces: computational demands and clinical needs meet basic neuroscience. Trends Neurosci 26(6):329–334

Nagasaka Y, Shimoda K, Fujii N (2011) Multidimensional recording (MDR) and data sharing: an ecological open research and educational platform for neuroscience. PLoS One 6(7):e22561

Nicolelis MAL (2003) Brain-machine interfaces to restore motor function and probe neural circuits. Nat Rev Neurosci 4(5):417–422

Onton J, Makeig S (2009) High-frequency broadband modulations of electroencephalographic spectra. Front Hum Neurosci 3:61

Patil PG, Turner DA (2008) The development of brain-machine interface neuroprosthetic devices. Neurotherapeutics 5(1):137–146

Quenouille MH (1949) Approximate tests of correlation in time-series. J R Stat Soc Series B (Methodol) 11(1):68–84

Ryali S et al (2010) Sparse logistic regression for whole-brain classification of fMRI data. Neuroimage 51(2):752–764

Schalk G et al (2007) Decoding two-dimensional movement trajectories using electrocorticographic signals in humans. J Neural Eng 4(3):264–275

Schelter B, Timmer J, Eichler M (2009) Assessing the strength of directed influences among neural signals using renormalized partial directed coherence. J Neurosci Methods 179(1):121–130

Schlögl A, Supp G (2006) Analyzing event-related EEG data with multivariate autoregressive parameters. Prog Brain Res 159:135–147

Schwartz AB et al (2006) Brain-controlled interfaces: movement restoration with neural prosthetics. Neuron 52(1):205–220

Schwarz G (1978) Estimating the dimension of a model. Ann Stat 6(2):461–464

Scott SH, Kalaska JF (1997) Reaching movements with similar hand paths but different arm orientations. I. Activity of individual cells in motor cortex. J Neurophysiol 77(2):826–852

Seth AK (2010) A MATLAB toolbox for Granger causal connectivity analysis. J Neurosci Methods 186(2):262–273

Szarowski D et al (2003) Brain responses to micro-machined silicon devices. Brain Res 983(1–2):23–35

Thakor N, Sherman D (1995) Wavelet (time-scale) analysis in biomedical signal processing. In: Bronzino J (ed) The biomedical engineering handbook. CRC, Boca Raton

Thomson DJ (1982) Spectrum estimation and harmonic analysis. Proc IEEE 70(9):1055–1096

Ting JA et al (2008) Variational Bayesian least squares: an application to brain-machine interface data. Neural Netw 21(8):1112–1131

Vetter RJ et al (2004) Chronic neural recording using silicon-substrate microelectrode arrays implanted in cerebral cortex. IEEE Trans Biomed Eng 51(6):896–904

Viventi J, Blanco J, Litt B (2010) Mining terabytes of submillimeter-resolution ECoG datasets for neurophysiologic biomarkers. In: Proceedings of the IEEE Engineering in Medicine and Biology Society Conference, Buenos Aires, 2010

Wiener N (1956) The theory of prediction. In: Beckenbach EF (ed) Modern mathematics for engineers. McGraw-Hill, New York, pp 165–190

Yamashita O et al (2008) Sparse estimation automatically selects voxels relevant for the decoding of fMRI activity patterns. Neuroimage 42(4):1414–1429

Zhang Y et al (2006) A cortical potential imaging study from simultaneous extra-and intracranial electrical recordings by means of the finite element method. Neuroimage 31(4):1513–1524

Zibulevsky M, Pearlmutter BA (2001) Blind source separation by sparse decomposition in a signal dictionary. Neural Comput 13(4):863–882

Ziehe A, Müller KR (1998) TDSEP-an efficient algorithm for blind separation using time structure. In: Niklasson L, Boden M, Ziemke T Proceedings of the 8th ICANN, perspectives in neural computing 1998. Springer, Berlin, pp 675–680

Chapter 4
Computational Analysis of Behavioural and Neural Data Through Bayesian Statistical Modelling

Raymond Wai Mun Chan and Fabrizio Gabbiani

Abstract Computational analysis of behavioural and neural data is nowadays an essential part of neuroethology, allowing an ever deeper understanding of how natural behaviour and neural activity are interrelated at the molecular, cellular and network level. The range of computational techniques applied in neuroethological research is currently so broad as to preclude an exhaustive survey in a succinct chapter. Here, we focus on a specific approach termed Bayesian statistical modelling that has proven to be a powerful method for relating neural activity to natural behavioural performance. As we illustrate in a specific example, this approach naturally dovetails with classic neural coding concepts such as population vector codes. It is also flexible enough to be applicable to a broad range of neuroethological questions.

Keywords Barn owl • Bayesian models • Interaural time difference • Maximum likelihood • Neural correlations • Population codes • Population vector • Sound localization

---

R.W.M. Chan
Department of Neuroscience, Baylor College of Medicine,
One Baylor Plaza, Houston, TX 77030, USA
e-mail: rchan@cns.bcm.edu

F. Gabbiani (✉)
Department of Neuroscience, Baylor College of Medicine,
One Baylor Plaza, Houston, TX 77030, USA

Department of Computational and Applied Mathematics,
Rice University, 6100 Main Street, Houston, TX 77005, USA
e-mail: gabbiani@bcm.edu

4.1 Introduction

During the past decades, neuroscience and neuroethology have experienced a dramatic increase in the availability of methods to analyse neural data. Yet, computational data analysis has long been an integral part of this area of research, as attested by many historical examples. Hodgkin and Huxley, for instance, used numerical integration of differential equations to study the propagation of action potentials in the giant squid axon. In another classical study, Katz and colleagues applied probability theory to derive the properties of quantal synaptic release at the neuromuscular junction. Recent progress in neural modelling has been in large part fueled by an exponential increase in computing power, the widespread availability of powerful numerical and simulation packages, such as MATLAB and NEURON, as well as the need to cope with increasingly complex neural data sets spanning multiple spatio-temporal scales. The interested reader will find a comprehensive treatment of modelling techniques in Gabbiani and Cox [2010], including many worked-out numerical and programming examples.

In this chapter, we focus on a specific topic that has attracted renewed attention and that is pertinent to neuroethology: Bayesian statistical modelling. The Bayesian framework allows the computational analysis of neural data in the context of the animal's environment using rigorous mathematical methods. In the following sections, we start with a brief introduction to Bayesian modelling before illustrating its use to analyze the neural coding of natural sounds in the barn owl. The figures of this chapter were generated using short MATLAB programs that are available online and will help the reader assimilate the material covered. The name of these programs is specified at the end of the figure legends using the notation: (name.m).

4.2 Bayesian Statistical Modelling

The Bayesian approach to statistics interprets probabilities as measures of belief instead of empirical frequencies for event occurrence (Doya et al. 2007, Hoff 2009). This framework centred on belief allows one to model decision making in a principled manner by (1) taking into account the sensory input experienced by an organism, (2) integrating previous information (e.g. memories or biases) and (3) deciding on an appropriate motor output, based on this information.

In the Bayesian framework, the state of the outside world may be conceived as a model indexed by a variable θ. In general, the variable θ will be multidimensional. The main task of the organism is to infer from sensory data, d, an estimate of the current state of the world, $\hat{\theta}$, so as to react with an appropriate motor output. The sensory data could for instance be the firing rate of sensory neurons activated in the current state of the world. Because the transduction of external stimuli into neural signals is noisy, due to both intrinsic and extraneous variability, and because the processing of neural signals is noisy as well, the sensory data d will usually be a random variable determined by θ and characterized by the conditional distribution

4 Bayesian Analysis of Behavioural and Neural Data

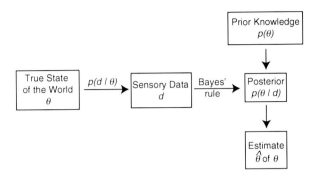

Fig. 4.1 Bayesian statistical models. In the Bayesian scheme, the observer gathers data from the outside world which is modelled based on parameters represented by θ. The data, d, is combined with prior knowledge on the parameters, θ, to infer a posterior probability distribution for the parameters using Bayes' rule [Eq. (4.1)]. This in turn allows to determine an estimate for the model parameters

$p(d|\theta)$. The main function on which decisions are based is the posterior distribution $p(\theta|d)$ which gives the conditional probability of the model parameter θ given the observed data d. To calculate this posterior distribution, we use Bayes' rule

$$p(\theta|d) = \frac{p(d|\theta)p(\theta)}{p(d)}, \qquad (4.1)$$

where $p(d|\theta)$ is the conditional probability of the data given the model parameter, θ (Fig. 4.1).

In the field of statistics, the probability distribution $p(d|\theta)$ is also called a generative model since it maps outside stimuli into sensory neural responses (in our context). The probability distribution of the model parameter θ, $p(\theta)$, is the prior distribution that is related to properties of the outside world. The experimenter will often be able to manipulate these priors. The distribution of responses irrespective of the stimuli (or of the current state of the world) is called the marginal distribution, $p(d)$. This marginal distribution normalizes Eq. (4.1) such that the integral of the posterior distribution over θ is unity. When we fix the data d and let θ vary, the generative model $p(d|\theta)$ becomes what is known as the likelihood function in statistics. One conventional method of estimating θ consists in selecting the value, $\hat{\theta}$, that maximizes the likelihood given the data. This decision rule is called "maximum likelihood". The analogous principle in the Bayesian case consists in selecting the maximum of the posterior distribution, or "maximum a posteriori" (MAP) estimate. Alternatively, another valid rule consists in computing the mean of the posterior distribution. The use of these decision rules will be illustrated in the following sections.

Often, knowing $p(d)$ is not necessary as we only need to know the dependence of the posterior on the model parameter, and $p(d)$ only acts as a normalizing constant. This is exploited by computing the product of the likelihood function and the prior distribution and ignoring the marginal distribution,

$$p(\theta \mid d) \propto p(d \mid \theta) p(\theta), \tag{4.2}$$

since $p(d)$ does not depend on θ. This last equation also makes clear that the Bayesian framework uses both the likelihood and the prior distribution of θ to arrive at an informed estimate $\hat{\theta}$. To render these general remarks more concrete, we turn to the example of sound localization in barn owls as recently described in Fischer and Peña [2011].

4.3 Sound Localization in Barn Owls

In the wild, owls use sound localization to detect and locate prey in the dead of night. Psychophysically, the time lag between a sound picked up in each ear but generated by a single source allows the owl to reconstruct the horizontal direction (or azimuth) to the source (Fig. 4.2a). A second and distinct cue, the interaural level difference allows the owl to reconstruct the elevation of the source but will not be considered further here (Konishi 2003). The time lag between sound arrival at both ears is called the interaural time difference (ITD), and is related to the azimuth direction of the sound source as shown in Fig. 4.2b. The horizontal axis shows the source direction centred on the owl's sagittal plane, while the vertical axis shows the corresponding ITD. This relationship is obtained by fitting the function

$$\text{ITD}(\theta) = A \sin(\omega \theta) \tag{4.3}$$

to head related transfer function data as a function of the source angle θ. Such a fit yields $A = 260\,\mu\text{s}$ and $\omega = 0.0143\,\text{rad}/°$ as the fitted parameters (Fischer and Peña 2011). From this graph, one notices immediately that the inverse mapping from ITD to source direction is not always one to one. Hence, the owl must somehow pick one of the possible states of the world consistent with the observed ITD. Ethologically, we know that it does so by biasing its choice to the one straight ahead (Hausmann et al. 2009, Knudsen et al. 1979). This bias can be quantified using Bayesian statistics.

We begin by asking what knowledge of the world the owl already has and what it wishes to know. In the sound localization problem, it knows approximately (see below) the ITD of the source, but wishes to know its associated direction θ. In Bayesian terms, we say that the owl wishes to infer the probability of θ given that it knows the ITD, or equivalently, the probability distribution of θ given the ITD, $p(\theta \mid \text{ITD})$. This is exactly the posterior distribution in Eq. (4.1) with ITD replacing d. In order to use Bayes' rule, we need a generative model and a prior distribution.

The sound reaching each ear may be corrupted by noise in the environment, like that caused by wind; in addition, the neural computation of ITD is noisy as well. Thus, we use

$$\text{ITD}(\theta) = A \sin(\omega \theta) + W, \tag{4.4}$$

4 Bayesian Analysis of Behavioural and Neural Data

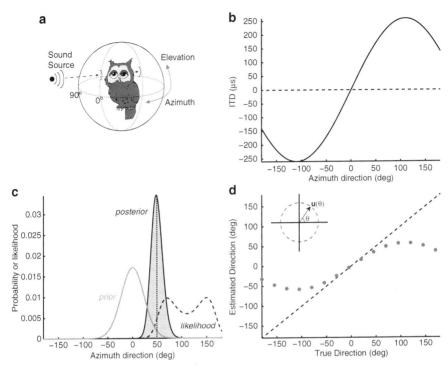

Fig. 4.2 Bayesian estimation of sound source. (**a**) Schematic illustration of the coordinate system used to describe sound direction, characterized by azimuth and elevation angles. A sound source with a horizontal direction or azimuth of 0° lies straight in front of the animal. A sound source with an azimuth different from 0° will arrive at a different time at the two ears. (**b**) Model of ITD as a function of azimuth fitted from experimental data. The *dashed horizontal line* indicates zero ITD. (**c**) The likelihood function or generative model for the θ, $p(\text{ITD}|\theta)$, is illustrated as the *dashed black line* (ITD = 218.9 ms). The *grey line* is the prior distribution expected by the owl, while the *black line* and *grey area curve* is the posterior probability distribution. The *dotted black line* represents the owl's sound source direction estimate ($\hat{\theta} = 49.7°$). (**d**) Estimated azimuth as a function of true (presented) azimuth based on the model described in the main text. The *dotted line* denotes model performance while the *dashed line* denotes ideal performance. Note the bias towards central positions exhibited by the model. The *upper left inset* illustrates the relation between the azimuthal angle θ and its associated unit vector $\mathbf{u}(\theta)$ on the unit circle (*dashed*) (bayesian.m)

where W is a Gaussian random variable with zero mean and standard deviation $\sigma_g = 41.2\,\mu s$. This gives the generative model

$$p(\text{ITD} | \theta) = \frac{1}{\sqrt{2\pi}\sigma_g} e^{-(\text{ITD} - A\sin(\omega\theta))^2 / (2\sigma_g^2)}. \tag{4.5}$$

The corresponding likelihood function is illustrated as the dashed curve of Fig. 4.2c for a specific value of the ITD. Notice that it is bimodal with two identical peaks. Hence the owl cannot simply select from this model a single most likely θ, according to the usual "maximum likelihood" principle.

A bias for one peak over the other has to be introduced to make a unique choice. This bias is derived from the prior distribution for the model parameter, θ, which we model as a Gaussian distribution of the form

$$p(\theta) = \frac{1}{\sqrt{2\pi}\sigma_p} e^{-\theta^2/(2\sigma_p^2)}, \quad (4.6)$$

where $\sigma_p = 23.3°$ is the standard deviation. In this case, the mean is zero which reflects the owl's strong bias for sound sources at the front, while tending to ignore possible sources from the sides. The prior distribution is shown in Fig. 4.2c as the grey curve. Note that θ is a valid angle on the unit circle only when $\theta \in (-180°, 180°]$, and we make the approximation that the probability mass is negligible outside these bounds. This prior is consistent with the known interactions of barn owls and their potential preys (Edut and Eilam 2004).

Applying Eqs. (4.5) and (4.6) to Eq. (4.2), we can compute the shape of the posterior distribution, shown as the grey area curve in Fig. 4.2c. With the posterior distribution, the owl can ask how probable the various source directions θ are given a measured ITD and use this information to make a behaviourally relevant choice. Making this choice involves using a decision rule to reduce a distribution over source directions $p(\theta|\text{ITD})$ to a single estimated source direction $\hat{\theta}$. One decision rule that is consistent with the behavioural data is to take an average of unit vectors weighted by their posterior probability of the form

$$\hat{\theta}(\text{ITD}) = \int \mathbf{u}(\theta) p(\theta | \text{ITD}) d\theta, \quad (4.7)$$

where $\mathbf{u}(\theta)$ is the two-dimensional unit vector for each angle (Fig. 4.2d, inset) and the integral is taken over the unit circle. This is also referred to as the circular mean. The result of this estimation procedure for a number of azimuth directions is given by the black points in Fig. 4.2d. As a reference, the black dashed line represents the perfect estimation. Note that the algorithm exhibits a bias towards central positions, that is, it tends to underestimate the true azimuth direction when the source is positioned at eccentric positions. This bias has been shown to exist in the barn owl by means of behavioural experiments. The good agreement between Eq. (4.7) and experimental data suggests that this equation may be implemented neurally, a topic we address in the following section.

4.4 Neural Encoding and Population Vector Decoding

We next ask how sound source localization in the barn owl is implemented by a population of neurons. One approach consists in building a model of the encoding process and then decoding the resulting neural activity using a population vector (PV). The population vector decoding method was pioneered more than 20 years

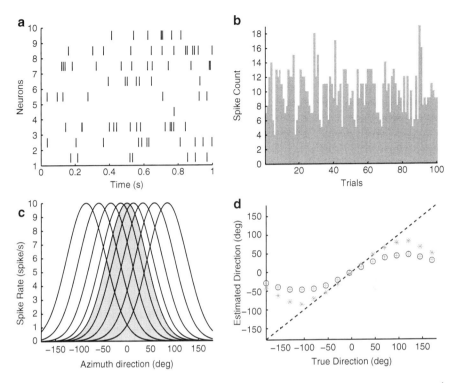

Fig. 4.3 Population vector (PV) estimate of sound source. (**a**) Raster plots of ten Poisson neurons' spike trains with a spike rate of 10 spike/s over a 1 s trial. (**b**) Histogram of a single Poisson spiking neuron' spike count across 100 trials (1 s long; 10 spike/s). (**c**) *Tuning curves* of nine neurons from the neural population encoding model. The *grey tuning curve* has a preferred direction of 0° for the sound source direction, with a peak firing rate of 10 spike/s. The preferred direction of the population is normally distributed around 0° [see Eq. (4.10)]. (**d**) Estimated azimuth as a function of true (presented) azimuth based on the PV and the probabilistic population code (PPC, see Sect. 4.5). The *grey crossed curve* represents the PV estimate while the *grey starred curve* represents the estimate obtained from the PPC (see Sect. 4.5). The *dashed line* denotes ideal performance. The *black circled curve* represents the estimate obtained from the PV when the neuron's firing rates are correlated ($\rho=0.5$, Sect. 4.6). Note that the bias towards central positions exhibited by the PV estimate is similar to that obtained from the PPC Bayesian Model (nnetwork.m)

ago in the superior colliculus and motor cortex to decode eye and arm movements from experimentally determined neuronal firing rates. Figure 4.3a shows the activity of ten Poisson neurons with spike rates of 10 spikes/s in a single trial (1 s long). For such Poisson neurons, the distribution of the spike count, k, in each 1 s bin has the form,

$$p(k) = \frac{\lambda^k}{k!}e^{-\lambda}, \tag{4.8}$$

where $\lambda = 10$ is the mean, as well as the variance (Ross 2007). The histogram of 100 trials from a single Poisson neuron with a rate of 10 spikes/s is shown in Fig. 4.3b. To test if such a spike histogram from an unknown distribution can be approximated by a Poisson spiking neuron, one can as a first step check if the ratio of the spike count variance and its mean is close to one. This ratio is called the Fano factor (Gabbiani and Cox 2010).

Sensory neurons change their spiking rate based on the specific external stimuli presented to an animal. This change in spike rate can be quantified using a tuning curve as shown by the grey area curve in Fig. 4.3c. The horizontal axis is the parameter of the stimulus, in our example the sound source direction, while the vertical axis is the average firing rate of the neuron responding to that stimulus parameter.

For our Poisson neuron, this would correspond to a 0° sound source and coincides with the peak firing rate of the neuron. We model the tuning curve with the form

$$r_n(\text{ITD}) = r_{\max} e^{-(\text{ITD} - A\sin(\omega\theta_n))^2/(2\sigma_g^2)}, \quad n = 1,\ldots,N, \tag{4.9}$$

where $r_{\max}$ is the peak firing rate and θ_n is the nth neuron's preferred direction. For our Poisson neuron, this would be 10 spikes/s and 0° respectively. The parameters A, ω and σ_g are the same as in Eq. (4.5); hence, Eq. (4.9) is proportional to Eq. (4.5). We shall assume that the population of neurons responsible for sound localization is homogeneous except that neurons have varying preferred directions, θ_n, as shown by the black curves in Fig. 4.3c.

The N neurons in the population have preferred direction θ_n sampled from the distribution

$$p(\theta) = \frac{1}{\sqrt{2\pi}\sigma_p} e^{-\theta^2/(2\sigma_p^2)}, \tag{4.10}$$

which is exactly the same as the prior in Eq. (4.6). Using the neuronal population vector of the form

$$\hat{\theta}(\theta) = \frac{1}{N}\sum_{n=1}^{N}\mathbf{u}(\theta_n)k_n \tag{4.11}$$

to decode the estimated sound source direction, we get results shown by the grey crossed curve in Fig. 4.3d. In Eq. (4.11), θ is the true sound source, k_n is the firing rate from a single trial of neuron n with tuning function given in Eq. (4.9) and $N=400$ is the number of neurons.

Notice that the curve looks strikingly similar to Fig. 4.2d. This is no coincidence, as our neural implementation can be shown to converge to the Bayesian estimate as the number of neurons $N \to \infty$ (Fischer and Peña 2011). Note also that in Eq. (4.9), $A\sin(\omega\theta)$ is the mean ITD, according to Eq. (4.4). Thus, a simple averaging mechanism is able to account for the behavioural data, based on the firing rate of a neuronal population tuned to ITD in a similar manner as barn owl neurons.

4.5 Probabilistic Population Codes

The PV is not the only method for decoding sensory responses from a population of neurons. An alternative scheme is based on a probabilistic population code (PPC; Ma et al. 2006). The PPC assumes that neuronal populations encode probability distributions through their joint firing rate tuning curves. As a result, the entire tuning curve of the neuronal population and not just the preferred direction is used in the decoding process. Let $\mathbf{k} = (k_1, k_2, ..., k_N)$ represent the response in a single trial of N neurons to a fixed sound source direction θ. The posterior distribution has the form

$$p(\theta \mid \mathbf{k}) \propto p(\mathbf{k} \mid \theta) p(\theta), \tag{4.12}$$

where $p(\theta)$ is the same as in Eq. (4.6) and $p(\mathbf{k}|\theta)$ is a distribution which models the probability of a neuronal response given the stimulus.

If we assume that the neurons representing $p(\mathbf{k}|\theta)$ are independent and Poisson, then the probabilistic population code for the distribution $p(\mathbf{k}|\theta)$ has the form

$$p(\mathbf{k} \mid \theta) = \prod_{n=1}^{N} \frac{r_n(\theta)^{k_n}}{k_n!} e^{-r_n(\theta)}, \tag{4.13}$$

where k_n is the response of neuron n and $r_n(\theta)$ is its tuning function. We model the tuning functions similarly as in Sect. 4.4, $r_n(\theta) = r_n(\text{ITD}(\theta))$ using Eqs. (4.9) and (4.3), but with a uniform distribution of preferred directions over the unit circle instead of being normally distributed. Note that the right-hand side of Eq. (4.13) is formed by taking products of Eq. (4.8) with k_i replacing k and $r_i(\theta)$ replacing λ, since we assume independent Poisson neurons.

Based on this probabilistic population code, an alternative decision rule to averaging over unit vectors is to pick the azimuth that maximizes the posterior probability. This rule is called the maximum a posteriori probability (MAP) rule and has the form

$$\hat{\theta}(\mathbf{k}) = arg_\theta \max p(\theta \mid \mathbf{k}).$$

The result of this estimation method based on the probabilistic population code is given in Fig. 4.3d as the grey starred curve. It is significantly different from the population vector result and does not match the behavioural data very well. On the other hand, a probabilistic population code has been successfully used to explain the sensory integration of visual and vestibular cues in neurons of the monkey visual cortex using a slightly different decoding mechanism (Fetsch et al. 2011).

4.6 Correlated Tuning Curves

In actual neural networks, the trial by trial firing rates of neurons may be correlated with each other (Averbeck et al. 2006, Ecker et al. 2010). To model these correlations, we assume that our N neurons have the same tuning curves and distributions of preferred directions as in Eqs. (4.9) and (4.10). In addition, we assume that their firing rates on a single trial are drawn from the multinormal probability distribution of the form

$$p(\mathbf{k} \mid \text{ITD}) = \frac{1}{(2\pi)^{1/(2N)} |\Sigma_g|^{1/2}} e^{-(\mathbf{k}-\mathbf{r}(\text{ITD}))' \Sigma_g^{-1} (\mathbf{k}-\mathbf{r}(\text{ITD}))/2} \qquad (4.14)$$

(Anderson 2003). In this equation, $\mathbf{r}(\text{ITD}) = (r_1(\text{ITD}),\ldots,r_N(\text{ITD}))'$ is the mean firing rate of the N neurons given the ITD, and $\mathbf{v}'$ denotes the transpose of vector $\mathbf{v}$. The covariance matrix is represented by Σ_g and its determinant by $|\Sigma_g|$. If the covariance matrix is given by $\Sigma_g = (\Sigma_{ij})$, with

$$\Sigma_{ij} = \sqrt{r_i(\text{ITD}) r_j(\text{ITD})} \delta_{ij}, \quad i,j = 1,\ldots,N,$$

and $\delta_{ij}=1$ when $i=j$, while $\delta_{ij}=0$ when $i \neq j$, we get an uncorrelated multinormal distribution. Because each neuron's firing rate variance is proportional to its mean firing rate, this formulation is close to that of Sect. 4.4 using Poisson neurons (mean equal to variance). If the covariance matrix elements have the form

$$\Sigma_{ij} = \sqrt{r_i(\text{ITD}) r_j(\text{ITD})} (\delta_{ij} + \rho(1-\delta_{ij})), \qquad (4.15)$$

where $\rho \in [0,1)$ is the correlation coefficient, we have introduced correlations of magnitude ρ into all pairs of neurons' firing rates.

When applying the PV, we see no difference in sound source estimates from independent neurons. This is illustrated in Fig. 4.3d by the black circled curve, which is exactly overlapping with the grey crossed curve obtained from independent neurons, in spite of sizable correlations between single neurons' firing rates ($\rho = 0.5$). Thus, neuronal correlations do not affect the results exposed in the previous sections. Intuitively, this may be understood from the fact that the direction of the PV will not be changed by correlated noise, if the noise scales uniformly with the mean firing rate of the neurons, as implemented by Eq. (4.15). In conclusion, Bayesian statistical modelling is a computational analysis technique that can provide insight in the coding of sensory information from a neuroethological perspective, as illustrated in this chapter.

References

Anderson TW (2003) An introduction to multivariate statistical analysis, 3rd edn Wiley, Hoboken

Averbeck BB, Latham PE, Pouget A (2006) Neural correlations, population coding and computation Nat Rev Neurosci 7(5):358–366

Doya K, Ishii I, Pouget A, Rao R (eds) (2007) Bayesian brain: probabilistic approaches to neural coding The MIT Press, Cambridge

Ecker AS, Berens P, Keliris Ga, Bethge M, Logothetis NK, Tolias AS (2010) Decorrelated neuronal firing in cortical microcircuits Science 327(5965):584–587

Edut S, Eilam D (2004) Protean behavior under barn-owl attack: voles alternate between freezing and fleeing and spiny mice flee in alternating patterns Behav Brain Res 155(2):207–216

Fetsch CR, Pouget A, DeAngelis GGC, Angelaki DDE (2011) Neural correlates of reliability-based cue weighting during multisensory integration Nat Neurosci 15(1):1–18

Fischer BJ, Peña JL (2011) Owl's behavior and neural representation predicted by Bayesian inference Nat Neurosci 14(8):1061–1066

Gabbiani F, Cox SJ (2010) Mathematics for neuroscientists Academic Press, San Diego

Hausmann L, Von Campenhausen M, Endler F, Singheiser M, Wagner H (2009) Improvements of sound localization abilities by the facial ruff of the barn owl (Tyto alba) as demonstrated by virtual ruff removal PLoS One 4(11):e7721

Hoff PD (2009) A first course in Bayesian statistical methods Springer texts in statistics Springer, Dordrecht

Knudsen EI, Blasdel GG, Konishi M (1979) Sound localization by the barn owl (*Tyto alba*) measured with the search coil technique J Comp Physiol 133(1):1–11

Konishi M (2003) Coding of auditory space Annu Rev Neurosci 26:31–55

Ma WJ, Beck JM, Latham PE, Pouget A (2006) Bayesian inference with probabilistic population codes Nat Neurosci 9(11):1432–1438

Ross SM (2007) Introduction to probability models, 10th edn Academic Press, San Diego

Part III
Optical Recording Techniques

Chapter 5
In Vivo Ca^{2+} Imaging of Neuronal Activity

Hiroto Ogawa and John P. Miller

Abstract Optical recording that provides both anatomical and physiological data has become an essential research technique for neuroethological studies. In particular, Ca^{2+} imaging is one of the most popular and useful methods for visualization of spatiotemporal dynamics of neuronal activity. Because Ca^{2+} is involved in so many fundamental neuronal signaling functions, including transmitter release and induction of synaptic plasticity, Ca^{2+} imaging can yield information that is crucial for a thorough understanding of these processes. In this chapter, we summarize aspects of Ca^{2+}-sensitive dyes that must be considered during the selection of an appropriate indicator for the specific question being investigated. We also discuss the development of dye-loading protocols, experimental designs, and optical system configurations that are required to enable the effective use of these Ca^{2+}-sensitive indicators. As an example application, we demonstrate how Ca^{2+} imaging of the cricket cercal sensory system in vivo has enabled us to monitor pre- and postsynaptic activity simultaneously on specific dendrites of an identified neuron.

Keywords Ca^{2+} imaging • Confocal microscope • Fluorescent Ca^{2+}-sensitive dye • Imaging device • In vivo imaging • Ratiometric imaging • Two-photon microscope

---

H. Ogawa (✉)
Department of Biological Science, Faculty of Science, Hokkaido University,
Kita 10, Nishi 8, Kita-ku, Sapporo 060-0810, Japan
e-mail: hogawa@sci.hokudai.ac.jp

J.P. Miller
Center for Computational Biology, Montana State University,
Bozeman, MT 59717, USA

5.1 Introduction

Optical recording is a powerful technique in neuroscience research, which enables us to simultaneously acquire data about the dendritic architecture of individual neurons, the multi-neuronal synaptic circuitry involving those neurons, and the time-resolved physiological response properties of those neurons. Rapid progress in computer technology, along with significant improvement of various types of imaging devices and microscopes, has driven substantial advances in optical recording techniques. At present, it is possible to visualize the neural activity not only in cultured cells and brain slices in vitro but also in the brains of intact, behaving animals in vivo. In neuroethological studies, optical recording methods have provided a number of important findings on the neural basis for various types of animal behavior. To get definitive and reliable data, however, pretreatment methods such as dye loading and sample preparation must be considered and carried out very carefully, and the imaging system and peripheral equipment must be configured in a manner that will support acquisition of data with the required spatial and temporal resolution.

In many cases, visualization of neural activity is mediated by specific probes that change their fluorescence or absorbance depending on membrane potential, cytosolic calcium concentration, and/or pH. Typically the synthetic organic dye is loaded into the targeted cells prior to the experiment, though it is now becoming common for a genetically encoded probe protein to be expressed in transgenic animals (see Chap. 7). It is also possible to visualize the neural activity as an intrinsic optical signal without probe loading, but fluorescent probes achieve much better signal-to-noise ratio and higher temporal resolution. In this chapter, we address Ca^{2+}-imaging techniques which use organic dyes having high and stable enough sensitivity to detect single action potentials or subthreshold synaptic activity. In particular, we will focus on in vivo Ca^{2+}-imaging techniques and describe points of concern for setting up the measurement instruments and pretreatments in a manner that are optimized for the intended research objectives. Details of more advanced Ca^{2+}-imaging methods are reviewed by Grienberger and Konnerth (Grienberger and Konnerth 2012).

5.2 Principle of Fluorescent Ca^{2+}-Sensitive Dyes

Organic Ca^{2+} indicators were developed in the 1980s by Tsien and his colleagues (Tsien 1980). All of Tsien's fluorescent Ca^{2+} indicators are based on a fluorophore combined with BAPTA, which retains high and stable selectivity for Ca^{2+} in the neutral to weak-alkaline pH range and has a fast time constant for Ca^{2+} binding. The Ca^{2+} indicators alter their spectral properties depending on the change in cytosolic Ca^{2+} concentration ($[Ca^{2+}]_i$) associated with membrane depolarization. In general, the Ca^{2+}-sensitive fluorescent dye is loaded into nerve cells using any of a variety of staining methods, and changes in the specific fluorescent intensity of the dye are measured with an imaging device such as a camera or photomultiplier mounted on a microscope.

The primary determining factor for selection of an appropriate fluorescent probe for a particular experimental situation is its affinity for Ca^{2+}, which is indicated by its dissociation constant (K_d). A wide variety of organic Ca^{2+} probes are now available having sensitivities to $[Ca^{2+}]_i$ which range from <50 nM to >50 μM in K_d values. In a typical nerve cell, local $[Ca^{2+}]_i$ can increase by a factor of between 10 and 100 in response to action potential generation. Therefore, if an indicator with very high affinity is used for imaging of neurons that fire spontaneously at high frequency, the indicator's fluorescent signal could be saturated instantly and would not monitor changes in the neural activity reliably. In such a case, an indicator with lower affinity would be chosen. On the other hand, if the experimental goal is to image Ca^{2+} influx from ligand-gated channels activated by synaptic inputs or Ca^{2+} transients evoked by one or a few action potentials, then indicators with a relatively high affinity of about 200 nM in K_d value are suitable. Thus, even within a single type of nerve cell, different indicators with different affinities would be chosen to image Ca^{2+} signals associated with processes going on at different spatial and temporal scales.

Another important consideration for the selection of an appropriate Ca^{2+} indicator is how the spectral characteristics of the fluorescent dye are changed by Ca^{2+} binding. The fluorescent indicators are broadly classified into singlemetric and ratiometric dyes (Fig. 5.1). The absorbance or fluorescence intensity of a singlemetric dye is changed depending on $[Ca^{2+}]_i$ without a shift in its emission spectrum. Typical singlemetric Ca^{2+} indicators, including fluo-3, fluo-4, Calcium Green, and Oregon Green 488 BAPTA, show a relative increase in the emission fluorescence with increase in $[Ca^{2+}]_i$. The behavior of fluo-3, a popular singlemetric indicator, is shown in Fig. 5.1a. Each plot in this family of curves shows the intensity of light emitted by the dye as a function of frequency, caused by illumination of the dye by light at 488 nm. Each different plot in the figure corresponds to a different $[Ca^{2+}]_i$. By monitoring changes in the total amount of light emitted from the sample within this range of 500–600 nm, the experimentor obtains a direct measurement of changes in $[Ca^{2+}]_i$. Since most of the singlemetric fluorescent dyes are all excited by visible light, Ca^{2+} imaging with these dyes requires a relatively simple optical system and can be achieved with a standard confocal microscope. Further, monitoring the fluorescence at only a single wavelength enables high-speed imaging of neural activity. However, singlemetric imaging in which $[Ca^{2+}]_i$ change is expressed as a relative fluorescence change with respect to the initial value ($\Delta F/F_0$) does not allow an effective cancellation of artifactual variation in the fluorescence signals resulting from photobleaching of the dye, fluctuation of the excitation light intensity, or movement of the sample. Therefore, in vivo Ca^{2+} imaging in awake animals using singlemetric indicators is extremely problematic and typically requires considerable ingenuity to eliminate the optical noise (including, e.g., stabilization of the neural tissue during recording and off-line post-processing using sophisticated software).

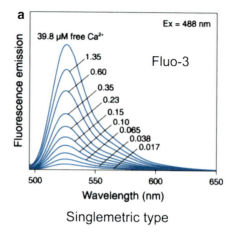
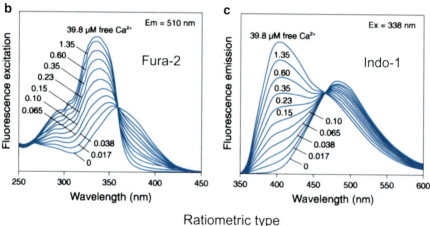

Fig. 5.1 The fluorescence emission and excitation spectra of three different organic synthetic Ca^{2+} dyes in solutions containing 0–39.8 μM free Ca^{2+}. (**a**) The emission at 530 nm of fluo-3, which is a typical singlemetric dye, is dependent on free Ca^{2+} concentration. The spectrum for the Ca^{2+}-free solution is indistinguishable from the baseline. (**b**) In Fura-2, a ratiometric dye, the ratio of 510 nm emission emitted by excitation at 340 and 380 nm wavelengths can be used to determine the concentration of Ca^{2+}. (**c**) Indo-1 is excited by 338 nm light. The peak of the emission fluorescence shifts from 400 to 480 nm with Ca^{2+} binding. The ratio of the fluorescence at 480 nm to at 400 nm indicates Ca^{2+} concentration (From The Molecular Probes Handbook)

Some of these technical limitations associated with singlemetric dyes can be circumvented through the use of ratiometric dyes, although at the expense of requiring a more complicated experimental configuration. The fluorescence behavior of ratiometric dyes also depends on $[Ca^{2+}]_i$. However, rather than responding to a change in $[Ca^{2+}]_i$ with a simple change in fluorescence intensity across the entire absorption spectrum, ratiometric dyes demonstrate a much more complicated behavior: the shapes of the absorption spectra of these indicators depends on $[Ca^{2+}]_i$.

As shown in figure panels 1B and 1C, the absorption spectra "shift" when $[Ca^{2+}]_i$ changes. These more complicated response characteristics enable an internally calibrated measurement of $[Ca^{2+}]_i$ that circumvents the problems listed in the preceding paragraph for singlemetric dyes. To achieve this independence from artifact, the experimentor must record a shift in the absorption spectrum of the indicator and thus must record two independent measurements of the indicator's fluorescence for each determination of $[Ca^{2+}]_i$.

To illustrate, consider Fura-2, which is one of the most popular ratiometric Ca^{2+} indicators. Figure 5.1b shows the excitation spectra of Fura-2 in the presence of different internal calcium concentrations. Note that these plots are very different from those shown in Fig. 5.1a: in panel a, each plot shows the intensity of light emitted by the dye as a function of frequency, caused by illumination of the dye by light at 488 nm. In panel b, each curve shows the excitation spectrum, not the emission spectrum: each curve plots the amplitude of fluorescence emitted by the indicator at a wavelength of 510 nm for a range of different excitation wavelengths between 250 and 450 nm. Each blue curve is the excitation spectrum at a different $[Ca^{2+}]_i$. During an experiment using Fura-2, the experimentor monitors the dye's fluorescence emitted at 510 nm in response to sequential illumination with light at two different wavelengths. The amplitude of the 510 nm fluorescence excited by 340 nm light increases with increasing $[Ca^{2+}]_i$, while the amplitude of the 510 nm fluorescence excited by 380 nm light decreases with increasing $[Ca^{2+}]_i$ (Fig. 5.1b). When two excitation beams with 340 and 380 nm wavelengths are delivered alternatively, the ratio of fluorescence intensity emitted at 510 nm by each excitation beam (F_{340}/F_{380}) can thus be used to calculate the change in $[Ca^{2+}]_i$.

Since the fluorescent images of Fura-2 excited at 340 and 380 nm are not acquired simultaneously, however, it becomes problematic to monitor neural activity at high temporal resolution using that indicator. The data acquisition rate is limited by the necessary exposure time at each of the two wavelengths. If a higher temporal resolution is required than can be achieved with Fura-2, another type of indicator may solve the problem. An example of such a one-excitation/two-emission type of ratiometric dye, Indo-1, is illustrated in Fig. 5.1c. This panel shows a family of emission spectra, as in panel A: each plot shows the intensity of light emitted by Indo-1 as a function of frequency, caused by illumination of the indicator by light at 338 nm. Each different plot in the figure corresponds to a different $[Ca^{2+}]_i$. Note that the peak of the fluorescence spectrum to excitation by 338 nm light shifts from 400 to 480 nm with increased Ca^{2+} binding (Fig. 5.1c). Thus, the ratio of the fluorescence intensity at 480 nm to that at 400 nm (F_{480}/F_{400}) indicates $[Ca^{2+}]_i$. For simultaneous measurement of the fluorescence intensities at 400 and 480 nm, the emission light is spectrally dispersed by a dichroic mirror, and two fluorescent images are divided with a special optical system (e.g., W-View System by Hamamatsu Photonics, DualView System by PHOTOMETRICS). Two simultaneous images are acquired with two cameras or onto the right and left halves of the imaging area of a single camera. The elimination of time lag in the acquisition of two images enables high-speed ratiometric imaging. FRET-based genetic probes such as "Cameleon" are another type of one-excitation/two-emission ratiometric indicator.

Most synthetic organic probes for ratiometric imaging are excited with UV light, which can be damaging to cells and requires a specialized laser for confocal imaging systems. Unfortunately, there are no available ratiometric Ca^{2+} sensors that have absorption within the visible range. However, it is possible to achieve effective ratiometric measurements in the visible-light range using two singlemetric probes like fluo-3 (or fluo-4) and Fura Red (Lipp and Niggli 1993; Speier et al. 2008) in the same cell(s). Both Fura Red and fluo-3 can be excited at the same visible wavelength (488 nm) but have different $[Ca^{2+}]_i$ dependencies. Therefore, if both dyes are simultaneously loaded into target cells, and two fluorescence images are recorded at 530 nm (fluo-3) and at 680 nm (Fura Red) using a spectral splitting system, then a change in $[Ca^{2+}]_i$ is indicated as the ratio of these fluorescence intensities (F_{530}/F_{680}). However, because the fluorescence of Fura Red is much weaker than that of the other visible-light excitable Ca^{2+} dyes, it is necessary to use a much higher concentration of the Fura Red than of the paired indicator.

In summary, in order to take full advantage of the potential benefits of Ca^{2+} indicators in studies of neural activity, investigators must pay careful attention to their requirements for temporal resolution and sensitivity, select an appropriate indicator, and then consider the design of the experimental protocols, sample preparation, and configuration of the optical recording system.

5.3 Experimental Protocols

5.3.1 *Dye Loading*

The synthetic organic Ca^{2+} probes are negatively charged and impermeable to cell membranes. There are two general methods for loading Ca^{2+} indicators into nerve cells: microinjection to a single cell and bulk loading to large populations of cells (Fig. 5.2). For loading into single cell, a grass micropipette or patch electrode filled with the Ca^{2+} dye of potassium salt is inserted into a target cell or applied for whole-cell recording, and the Ca^{2+} indicator is diffused directly into cytoplasm (Fig. 5.2a). When using a sharp microelectrode, the Ca^{2+} dye can be loaded electrophoretically with depolarizing current injection. The microinjection method for staining a single cell can provide a very bright image of the cell over the background, achieving a high signal-to-noise ratio for the optical recordings. This method is preferred for measurements requiring high spatial resolution, such as recording the Ca^{2+} transients in subcellular regions like dendritic branches or dendritic spines. Further, this method enables simultaneous measurement of $[Ca^{2+}]_i$ changes with electrophysiological recording of action potentials or synaptic currents. This approach has yielded significant findings related to synaptic integration in dendritic arborization.

There are several techniques for the bulk loading of the Ca^{2+} dye into larger populations of nerve cells. The most popular method for loading fluorescent indicators is by the application to the tissue of acetoxymethyl (AM) ester, which is a

5 In Vivo Ca²⁺ Imaging of Neuronal Activity

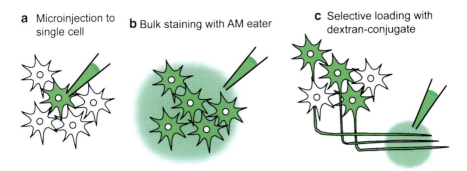

Fig. 5.2 Three classes of dye-loading methods. (**a**) For staining of single neuron, cell-impermeable Ca^{2+} indicator can be injected into the cell through a sharp or whole-cell patch electrode. (**b**) Bath application of AM ester of the Ca^{2+} indicator via a micropipette inserted into the nervous tissue is used for bulk staining of many neurons. (**c**) Uptake of the dextran-conjugated Ca^{2+} indicators into the axonal fibers running through a specific tract is capable of selective staining of projection neurons

cell-permeable form of the indicator developed by Tsien (Tsien 1981). The AM ester of the indicator is, itself, insensitive to Ca^{2+} and will not fluoresce outside of cells. Once inside the cell, however, the AM ester of the Ca^{2+} indicator is hydrolyzed by intracellular esterase, restoring the Ca^{2+} sensitivity and also making the indicator membrane impermeable so that it will not leak out. For staining cells in culture or in brain-slice preparations, the AM ester of the Ca^{2+} indicator is delivered along with a nonionic detergent, Pluronic F-127 or PowerLoad (Invitrogen), which are reagents that help disperse the indicator (Fig. 5.2b). To use the AM ester staining techniques to in vivo preparations, the loading solution is typically pressure injected directly into the brain or ganglion preparation (Garaschuk et al. 2006). However, there is considerable variability in the loading efficiency between different neuronal tissues, between different cell types within a single tissue sample, and between different animal species. This variability is due to differences in membrane permeability for the AM ester and in the endogenous intracellular esterase activity. Since the AM ester is loaded not only into neurons but also into glial cells, dual staining with Ca^{2+} indicator and astrocyte markers is often required to distinguish neuronal responses from Ca^{2+} signals originating from the glial cells (Garaschuk et al. 2006).

Dextran-conjugated Ca^{2+} indicators are useful for selective loading into axonal fibers that lie within specific nerve tracts. Dextrans are hydrophilic polysaccharides characterized by their high molecular weight, high water solubility, and low toxicity. Due to these properties, fluorescent dextrans have been used for anterograde and retrograde neuron tracing. Dextran-conjugated dyes loaded into the neurons are effectively transported by axonal trafficking and allow staining of a specific group of neurons (Fig. 5.2c). For example, a highly concentrated chip of dextran-conjugated Ca^{2+} indicator can be placed on a specific location within the brain and will be taken up by a very restricted set of axons in contact with that chip (Gelperin and Flores 1997; Delaney et al. 2001; Sachse and Galizia 2002).

Recently, electroporation techniques have been used to achieve bulk loading of Ca^{2+} indicators (Nagayama et al. 2007; Hovis et al. 2010; Bonnot et al. 2005). Following microinjection of the potassium salt or the dextran-conjugated fluorescent dye into the brain, an electric field is imposed across a population of cells to transiently introduce membrane pores through which the Ca^{2+} indicators can enter. This method is capable of loading the Ca^{2+} indicators into brain tissue that is inaccessible to the AM ester.

In summary, there are a variety of approaches for loading the Ca^{2+} indicators into the target cell or cells to be studied. The most effective method will depend upon the nature of the preparation under study, the specific questions to be addressed, and any technological limitations imposed by the recording method to be used.

5.3.2 Preparation for In Vivo Imaging

In vivo Ca^{2+} imaging of nerve cells in awake behaving animals is ideal for neuroethogical research but presents a host of technical difficulties beyond those associated with dye loading. Problems related to stabilization of the target tissue and accessibility of that tissue to a high-numerical-aperture microscope objective require careful consideration. To illustrate some of these considerations, we introduce three different types of preparations using the cricket nervous system for in vivo Ca^{2+} imaging.

The first type of preparation is an isolated preparation, in which an entire ganglion and connected sensory organs are removed from the animal's body. For our studies of the neuronal processing underlying the detection and analysis of air-current direction and dynamics in the cricket cercal sensory system, we dissect the terminal abdominal ganglion away from the abdomen, along with the pair of air-current-sensitive organs called cerci that send sensory input into the ganglion. The preparation is mounted in a glass chamber made of a 0.25-mm thick cover glass (Ogawa et al. 2006, 2008). This recording chamber is mounted on the stage of an inverted microscope configured for fluorescent imaging. We can observe clear images of dye-loaded neurons in this preparation, maintaining the neural circuits and the sensory organs in conditions that are similar to their natural in vivo configuration. By removing the standard condenser tube and transmitted light source from the microscope, which normally obstruct access to sample from above, this setup provides excellent accessibility of the preparation to electrodes and perfusion apparatus and has the advantage that the objective lens (placed below the stage) does not disturb the normal airflow stimulus across the sensory organs.

The second type of preparation is a "head-fixed" preparation, in which the nervous system is left within a restrained animal's body. The head-fixed preparation enables in vivo imaging experiments to be carried out in intact animals that contain a complete, intact central nervous system with fully functional sensory capabilities. Pioneer works of the in vivo Ca^{2+} imaging in the head-fixed preparation were performed in fly visual system, demonstrating dendritic Ca^{2+} accumulation evoked by visual motion stimuli (Borst and Egelhaaf 1992; Single and Borst 1998). At present, two-photon imaging experiments in head-fixed mice have successfully monitored cortical Ca^{2+} responses to whisker, odor, auditory, and visual stimuli (Sato et al. 2007; Wachowiak et al. 2004; Bandyopadhyay et al. 2010; Rothschild et al. 2010; Smith and Häusser 2010). A critical requirement for this preparation is the immobilization of the nervous tissue with respect to the objective lens. We have carried out optical recordings using awake crickets that were fixed on a silicone elastomer platform by insect pins, with their brain ganglion held up and immobilized using a stainless steel spoon. In a preliminary study using this preparation, we have optically recorded the odor-evoked Ca^{2+} signals in the calyx of the mushroom body in a conditioned cricket (Fig. 5.3; Ogawa and Oka 2008).

The most ideal preparation for neuroethological studies is a tethered preparation, which is designed for in vivo imaging of animals performing an actual behavior, like locomotion. This type of preparation is more difficult with respect to stabilization of the nervous tissue: it is important to only immobilize the nervous tissue to be studied under the microscope, without encumbering the animal's behavior. One novel imaging method has recently been developed to achieve this goal, through which a microendoscope made of narrow optical fiber bundles is penetrated into the brain of a freely moving mouse enabling dendritic Ca^{2+} imaging (Adelsberger et al. 2005; Murayama et al. 2007). Here we summarize a tethered preparation for in vivo imaging of the cricket during locomotion on a spherical treadmill. A U-shaped steel plate was slipped into the cervical segment between the head and thorax, and the cuticular epicranium was fixed to that plate with wax. One end of the plate was connected to a stainless steel rod, controlled via a micromanipulator, for precise positioning of the head. This head-tethered cricket was placed so that its legs were positioned on a treadmill in their natural arrangement, and fluorescent images of the brain ganglion were monitored under the microscope during the cricket's walking behavior. Using similar head-fixed preparations in mice and *Drosophila*, two-photon Ca^{2+} imaging has been performed while the animals were walking (Dombeck et al. 2007; Seelig et al. 2010). To eliminate optical noise resulting from vibration of the brain during the animals' movements, additional procedures were required involving a surgical fixture or filling with a silicone adhesive (Kwik-Sil, WPI).

5.3.3 *Imaging Devices and Data Analysis*

Various types of fluorescent microscopes have been developed over the last few decades, and most configurations have now been used to carry out Ca^{2+} imaging in neuroscience research. Likewise, multiple kinds of imaging devices have been developed and are available for imaging experiments. In general, Ca^{2+} imaging requires cameras with very high sensitivity. Current implementations typically use an electron-multiplying CCD (EM-CCD) or a complementary metal oxide semiconductor (CMOS) sensor mounted to a conventional fluorescent microscope. Recent advances in microscope and imaging technologies are providing progressively

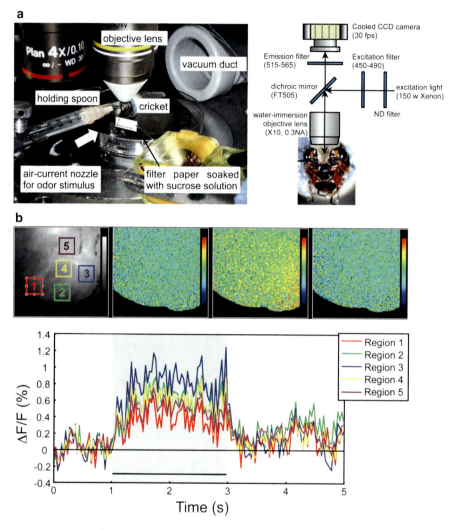

Fig. 5.3 In vivo Ca^{2+} imaging in the head-fixed cricket preparation. (**a**) Schematic of the experimental setup showing an immobilized cricket. (**b**) Ca^{2+} response in the calyx of the mushroom body to the preferred odor stimulus

brighter fluorescent images with improved signal-to-noise ratio. Regardless of the kind of imaging device used, however, the optical properties of the objective lens and optical band-pass filters are the most crucial determinants of image quality. A primary consideration for setting up a Ca^{2+}-imaging experimental system is, therefore, the selection of an objective lens with as large a numerical aperture (N.A.) as possible, adequate working distance, and a set of filters appropriate for transmitting the excitation and emission wavelengths for the selected Ca^{2+} indicator(s). An appropriate imaging device must also be selected that is compatible with the emission characteristics of the indicator(s), the optical configuration, and the required spatial and temporal resolutions. Depending on the staining and preparation of the tissue, optical configuration, and imaging device characteristics, it is currently possible to acquire images at rates of up to 1,000 frames/second (fps).

With conventional fluorescent microscopy, a high-pressure mercury lamp or a xenon lamp can provide stable excitation illumination over a wide range of wavelengths. The use of light-emitting diodes (LEDs) as illumination sources is a more recent innovation and has been found to have several advantages over conventional sources. An LED can supply illumination at only a single wavelength but offer a lot of advantages as excitation sources if an appropriate wavelength is available: the size of the source is very small, the fluctuation in intensity due to heat production is very low, the intensity is easily variable over a relatively wide range (reducing the need for neutral density filters), and the illumination can be switched on and off very rapidly, eliminating the need for shutters in the excitation light path.

Light scattering is one of the most serious problems that arise for in vivo brain imaging. The scattering problem is greatly reduced through the use of confocal laser-scanning microscopes (CLSMs), which are currently the most popular imaging systems used for high-resolution Ca^{2+} imaging. Inexpensive laser sources with excellent lifetimes have recently become available. Two-photon microscopy (TPM) has also become a leading-edge imaging technology in neuroscience. TPMs have profound advantages for in vivo imaging because they are capable of observing neuronal activity located at relatively deep layers in whole brain preparations (Helmchen and Denk 2005; Nemoto 2008; Svoboda and Yasuda 2006). However, CLSM and TPM have relatively lower temporal resolution than conventional camera-based devices for capturing images of large fields, because the operation of scanning the laser over an entire field of view is more time consuming than "snapping" an image of the same field with a conventional camera-based device. An effective way to mitigate this limitation to the temporal resolution is to operate the CLSM or TPM in "line-scanning mode": the investigator captures only a linear "transect" through the region of interest in the field of view, rather than collecting an image of the entire field of view. This line-scanning mode is typically used for measurement of Ca^{2+} transients evoked by single action potentials or EPSP. In this mode, however, it is impossible to record the two-dimensional patterns of Ca^{2+} signals. Nipkow-type spinning-disk microscopes are an effective intermediate solution: they substantially improve the time resolution for confocal imaging over conventional scanning laser configurations. This microscope provides a confocal effect by spinning a disk with microlens-embedded pinholes in the excitation light path. However, the image acquired with the Nipkow-disk microscopy is less intense than that obtained through conventional CLSM. Recently, the use of a high-sensitivity digital camera like the EM-CCD camera as the image acquisition device for Nipkow-disk microscopy has enabled much brighter confocal imaging at higher time resolution. Using this technology, investigators have detected Ca^{2+} transients evoked by individual action potentials at 150 μm tissue depth (Takahashi et al. 2010; Takahara et al. 2011).

Another very important aspect that must be considered during the configuration of an effective optical imaging system is the system integration: synchronization of the stimulator, electrophysiological recording apparatus, and all other peripheral equipment associated with real-time image capture. Generally, digital TTL signals are used for synchronizing the operation of image acquisition and temporal control of the onset and duration and frequency of stimulation. There is, however, a brief lag between the trigger of the TTL signal and the initiation of image acquisition within the computer. This time lag is inevitable and ranges from a few to dozens of milliseconds, depending on the frame rate. Therefore, it is necessary to design the circuitry for synchronization of the experimental components in a way that the time delay is minimized and constant between trials. In our system, a TTL signal exported from the imaging computer is used for triggering the stimulation and electrophysiological recordings and also receives feedback from the stimulus monitor.

Fairly sophisticated software is typically provided along with any commercially available advanced imaging device and is usually adequate for sequential image acquisition required for Ca^{2+} imaging. However, subsequent data analysis of the optical recordings often requires the development and application of specialized algorithms, and it is very typical for an investigator to use other more specialized research-oriented image-processing software packages like "Image J (NIH)" (Dreosti et al. 2009). As mentioned above, singlemetric imaging techniques yield signals that characterize changes in $[Ca^{2+}]_i$ as relative changes in the observed fluorescence, $\Delta F/F_0$, where F_0 is the fluorescence intensity of the first image (or the mean value of all images during the prestimulation period). For visualization of Ca^{2+} signals using pseudo-color images, it is important to configure the minimum and maximum values to capture the entire dynamic range of all images throughout the recording. If the fluorescence change is too small and/or signal-to-noise ratio is too low, several sequences of images must be acquired in response to repeated stimuli to enable signal averaging. However, signal averaging for noise reduction is not attainable for imaging of spontaneous or motor-related activity, because these activities occur with variable delay to the stimulus onset.

5.4 Application: Simultaneous Imaging of Pre- and Postsynaptic Neurons

In the final section of this chapter, we summarize a procedure we used for simultaneous Ca^{2+} imaging of pre- and postsynaptic neurons. For these experiments, we combined single-neuron staining of a postsynaptic neuron using microelectrode injection of one Ca^{2+} indicator and bulk staining of presynaptic fibers with the AM ester of a second Ca^{2+} indicator. We simultaneously monitored the optical signals from both indicators in different wavelengths using a dual-view optical system. This method enabled us to measure the pre- and postsynaptic activity on specific dendrites of the identified interneuron. We adapted this method to the cricket cercal sensory system (Ogawa et al. 2008).

The cercal sensory system in the cricket detects the direction, frequency, and velocity of air currents with great accuracy and precision. The receptor organs of this system consist of a pair of antenna-like appendages called cerci at the rear of cricket abdomen. Each cercus is covered with approximately 500 filiform hairs, each of which is innervated by a single mechanoreceptor neuron. The receptor neuron is tuned to air currents from a particular direction and exhibits a change in its firing rate in response to stimuli over the entire 360° range of stimulus directions (Landolfa and Miller 1995). Optical imaging of a population of the mechanosensory afferents stained with the AM ester of a Ca^{2+} indicator demonstrated that the direction of the air currents is represented by specific spatial patterns in the ensemble activities of the afferents (Ogawa et al. 2006). Identified giant interneurons (GIs) receive direct excitatory synaptic inputs from the mechanosensory afferents. The GIs are activated by air currents and also display differential sensitivity to variations in air-current direction (Jacobs et al. 1986; Miller et al. 1991; Theunissen et al. 1996). The directional tuning curves of the GIs are well described by a cosine function; that is, the GIs encode information about the stimulus direction proportional to their spiking activity. To describe decoding algorithm of the directional information from the population activities of sensory afferents to individual GIs, we measured the pre- and postsynaptic local Ca^{2+} responses to the air-current stimuli, on each GI's dendrite.

Two kinds of Ca^{2+} indicators with different fluorescent wavelengths were loaded to pre- and postsynaptic neurons, respectively. The sensory afferent fibers were stained with AM ester of Oregon Green 488 BAPTA-1 (OGB-1), while cell-impermeant Fura Red was injected into the single GI (Fig. 5.4a). For staining the sensory afferents, a solution containing AM ester of OGB-1 at a concentration of 0.05 % and dispersing reagent (Pluronic F-127, Invitrogen) at a concentration of 1 % was pressure injected into a cercal sensory nerve through a glass micropipette. Twelve hours after dye injection, the axon terminals of cercal sensory neurons were found to be stained with OGB-1. After staining of the afferents with OGB-1, 2 mM Fura Red tetrapotassium salt was iontophoretically injected into the GI for 5 min through a glass microelectrode, using a hyperpolarizing current of 3 nA. Fluorescent signals were viewed with an inverted microscope (Axiovert100, Zeiss). A xenon arc lamp (XBO 75 w, Zeiss) illumination with a stabilized power supply and 470/20 band-pass filter were used for excitation of the Ca^{2+} indicators, OGB-1, and Fura Red. For simultaneous measurement of pre- and postsynaptic Ca^{2+} signals, a fluorescent image passing through a FT510 dichroic mirror was divided into two images with *W-View* optics (Hamamatsu Photonics) by the following filter set: dichroic 590LP, emission 535/45 for the OGB-1 and 610/25 for Fura Red (Fig. 5.4b). The two separated images were simultaneously acquired side by side in the same frame with a digital cooled CCD camera (ORCA-ER, Hamamatsu Photonics) attached to the inverted microscope.

Using this optical system to separate the optical signals from OGB-1 and Fura Red, we simultaneously measured the Ca^{2+} responses to air-current stimuli in the GI's dendrites and in the afferent axon terminals that make synaptic connections onto that dendritic branch (Ogawa et al. 2008). It was observed that the air-current stimulus evoked a significant decrease in the fluorescence of Fura Red (610 nm

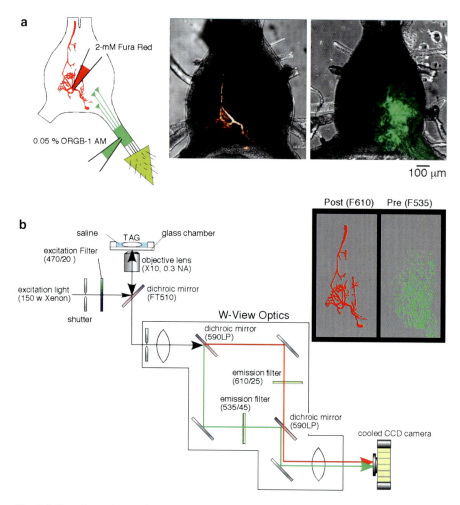

Fig. 5.4 Experimental setup for simultaneous imaging of Ca^{2+} signals in the pre- and postsynaptic neurons in the cricket cercal sensory system. (**a**) Diagrams showing the methods for selective loading of the different Ca^{2+} indicators. A solution containing OGB-1 AM was pressure injected through a glass micropipette into a cercal sensory nerve. The potassium salt of Fura Red was iontophoretically injected into the dendrites of the identified interneuron through a sharp electrode with a hyperpolarizing current. Right panels are superimposed displays of confocal images showing fluorescence of Fura Red injected into the interneuron 10-2 (*left image*) or OGB-1 loaded into the afferent axons of a left cercus (*right image*) over the transmitted light images of the terminal abdominal ganglion. (**b**) Diagrams of the optical splitting system for simultaneous monitoring two fluorescent wavelengths of OGB-1 and Fura Red. A fluorescent image was divided into two images by W-View optics with a set of dichroic mirrors and emission filters. Both images were acquired in the same frame side by side with a cooled CCD camera at the same time (modified from Ogawa et al. 2008)

5 In Vivo Ca²⁺ Imaging of Neuronal Activity

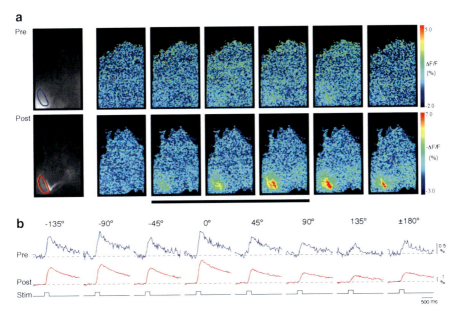

Fig. 5.5 Pre- and postsynaptic Ca²⁺ imaging of interneuron 10-3. (**a**) Ca²⁺ responses to air-current stimuli applied from the anterior orientation (0°). The left two (monochrome) images are raw pre-stimulus fluorescent images. A series of six images to the right of the baseline images are pseudo-colored images indicating the [Ca²⁺]ᵢ elevation in the sensory afferents (*upper*) and the interneuron 10-3 (*lower*). (**b**) The time courses of changes in ΔF/F at 535 nm wavelength (presynaptic) Ca²⁺ signals (*blue traces*) and ΔF/F at 610 nm wavelength (postsynaptic) Ca²⁺ signals (*magenta traces*). These traces were recorded in the dendritic branch shown as *ROIs* in the raw fluorescent images in (**a**). The air-current stimuli were applied from eight different orientations (modified from Ogawa et al. 2008)

wavelength), which indicates an elevation in [Ca²⁺]ᵢ at the dendritic region of the postsynaptic GI. Simultaneous measurement of light with 535 nm wavelength at the same recording area showed a fluorescence increase of OGB-1 indicating a rise in Ca²⁺ in the sensory afferents. We recorded the pre- and postsynaptic local responses to the air-current stimuli on each dendrite of the GIs. We applied air-current stimuli from eight different directions in the horizontal plane and examined the directional sensitivity of the Ca²⁺ responses of the sensory afferents and of the GI. Figure 5.5 shows typical responses to the air-current stimuli in the pre- and postsynaptic Ca²⁺ changes, which were measured at three different dendritic branches of the GI named 10-3. Presynaptic and postsynaptic Ca²⁺ responses showed directional tuning properties in their response amplitudes. Although the directional tuning properties of postsynaptic responses on individual dendrites varied from each other, the directional sensitivity of dendritic Ca²⁺ responses corresponded to those indicated by Ca²⁺ signals in presynaptic afferents arborizing on that dendrite. This similarity in the directional sensitivity between the pre- and postsynaptic Ca²⁺ responses suggests that the individual dendrite with a distinct tuning property in its Ca^{2+} response receives synaptic inputs from different subsets of the afferents having different sensitivities to direction.

Thus, simultaneous Ca^{2+} imaging of pre- and postsynaptic neurons has provided significant findings on the decoding algorithm for synaptic transmission and on the dendritic integration of synaptic inputs. Recent approaches using high-speed two-photon imaging are capable of visualizing and functionally mapping the synaptic inputs onto individual spines on the dendrites of cortical neurons (Jia et al. 2010; Chen et al. 2011; Varga et al. 2011). In neuroethological research, further development of in vivo Ca^{2+}-imaging techniques and the development of novel indicators, including genetic probes, will be amenable major advances in our understanding of the neuronal architecture and computations underlying animal behavior.

References

Adelsberger H, Garaschuk O, Konnerth A (2005) Cortical calcium waves in resting newborn mice. Nat Neurosci 8:988–990

Bandyopadhyay S, Shamma SA, Kanold PO (2010) Dichotomy of functional organization in the mouse auditory cortex. Nat Neurosci 13:361–368

Bonnot A, Mentis GZ, Skoch J, O'Donovan MJ (2005) Electroporation loading of calcium-sensitive dyes into the CNS. J Neurophysiol 93:1793–1808

Borst A, Egelhaaf M (1992) In vivo imaging of calcium accumulation in fly interneurons as elicited by visual motion stimulation. Proc Natl Acad Sci USA 89:4139–4143

Chen X, Leischner U, Rochefort NL, Nelken I, Konnerth A (2011) Functional mapping of single spines in cortical neurons in vivo. Nature 475:501–505

Delaney K, Davison I, Denk W (2001) Odour-evoked [Ca^{2+}] transients in mitral cell dendrites of frog olfactory glomeruli. Eur J Neurosci 13:1658–1672

Dreosti E, Odermatt B, Dorostkar MM, Lagnado L (2009) A genetically encoded reporter of synaptic activity in vivo. Nat Methods 6:883–889

Dombeck DA, Khabbaz AN, Collman F, Adelman TL, Tank DW (2007) Imaging large-scale neural activity with cellular resolution in awake, mobile mice. Neuron 56:43–57

Garaschuk O, Milos RI, Konnerth A (2006) Targeted bulk-loading of fluorescent indicators for two-photon brain imaging in vivo. Nat Protoc 1:380–386

Gelperin A, Flores J (1997) Vital staining from dye-coated microprobes identifies new olfactory interneurons for optical and electrical recording. J Neurosci Methods 72:97–108

Grienberger C, Konnerth A (2012) Imaging calcium in neurons. Neuron 73:862–885

Helmchen F, Denk W (2005) Deep tissue two-photon microscopy. Nat Methods 2:932–940

Hovis KR, Padmanabhan K, Urbana NN (2010) A simple method of in vitro electroporation allows visualization, recording, and calcium imaging of local neuronal circuits. J Neurosci Methods 19:1–10

Jacobs GA, Miller JP, Murphey RK (1986) Cellular mechanisms underlying directional sensitivity of an identified sensory interneuron. J Neurosci 6:2298–2311

Jia H, Rochefort NL, Chen X, Konnerth A (2010) Dendritic organization of sensory input to cortical neurons in vivo. Nature 464:1307–1312

Landolfa MA, Miller JP (1995) Stimulus–response properties of cricket cercal filiform receptors. J Comp Physiol A 177:749–757

Lipp P, Niggli E (1993) Ratiometric confocal Ca^{2+}-measurements with visible wavelength indicators in isolated cardiac myocytes. Cell Calcium 14:339–372

Miller JP, Jacobs GA, Theunissen FE (1991) Representation of sensory information in the cricket cercal sensory system. I. Response properties of the primary interneurons. J Neurophysiol 66:1680–1689

Murayama M, Pérez-Garci E, Lüscher HR, Larkum ME (2007) Fiberoptic system for recording dendritic calcium signals in layer 5 neocortical pyramidal cells in freely moving rats. J Neurophysiol 98:1791–1805

Nagayama S, Zeng S, Xiong W, Fletcher ML, Masurkar AV, Davis DJ, Pieribone VA, Chen WR (2007) In vivo simultaneous tracing and Ca^{2+} imaging of local neuronal circuits. Neuron 53:789–803

Nemoto T (2008) Living cell functions and morphology revealed by two-photon microscopy in intact neural and secretory organs. Mol Cells 26:113–120

Ogawa H, Cummins GI, Jacobs GA, Miller JP (2006) Visualization of ensemble activity patterns of mechanosensory afferents in the cricket cercal sensory system with calcium imaging. J Neurobiol 66:293–307

Ogawa H, Cummins GI, Jacobs GA, Oka K (2008) Dendritic design implements algorithm for extraction of sensory information. J Neurosci 28:4592–4603

Ogawa H, Oka K (2008) In vivo calcium imaging of mushroom body calyx in the tethered cricket for odor-taste conditioning. Program No. 362.20. Neuroscience Meeting Planner. Society for Neuroscience, Washington, DC, 2008

Rothschild G, Nelken I, Mizrahi A (2010) Functional organization and population dynamics in the mouse primary auditory cortex. Nat Neurosci 13:353–360

Sachse S, Galizia G (2002) Role of Inhibition for temporal and spatial odor representation in olfactory output neurons: a calcium imaging study. J Neurophysiol 87:1106–1117

Sato TR, Gray NW, Mainen ZF, Svoboda K (2007) The functional microarchitecture of the mouse barrel cortex. PLoS Biol 5:e189

Seelig JD, Chiappe ME, Lott GK, Dutta A, Osborne JE, Reiser MB, Jayaraman V (2010) Two-photon calcium imaging from head-fixed *Drosophila* during optomotor walking behavior. Nat Methods 7:535–540

Single S, Borst A (1998) Dendritic integration and its role in computing image velocity. Science 281:1848–1850

Smith SL, Häusser M (2010) Parallel processing of visual space by neighboring neurons in mouse visual cortex. Nat Neurosci 13:1144–1149

Speier S, Nyqvist D, Cabrera O, Yu J, Molano RD, Pileggi A, Moede T, Köhler M, Wilbertz J, Leibiger B, Ricordi C, Leibiger IB, Caicedo A, Berggren P (2008) Noninvasive in vivo imaging of pancreatic islet cell biology. Nat Med 14:574–578

Svoboda K, Yasuda R (2006) Principles of two-photon excitation microscopy and its applications to neuroscience. Neuron 50:823–839

Takahara Y, Matsuki N, Ikegaya Y (2011) Nipkow confocal imaging from deep brain tissues. J Integr Neurosci 10:121–129

Takahashi N, Sasaki T, Matsumoto W, Matsuki N, Ikegaya Y (2010) Circuit topology for synchronizing neurons in spontaneously active networks. Proc Natl Acad Sci USA 107:10244–10249

Theunissen F, Roddey JC, Stufflebeam S, Clague H, Miller JP (1996) Information theoretic analysis of dynamical encoding by four primary sensory interneurons in the cricket cercal system. J Neurophysiol 75:1345–1376

Tsien RY (1980) New calcium indicators and buffers with high selectivity against magnesium and protons: design, synthesis, and properties of prototype structures. Biochemistry 19:2396–2404

Tsien RY (1981) A non-disruptive technique for loading calcium buffers and indicators into cells. Nature 290:527–528

Varga Z, Jia H, Sakmann B, Konnerth A (2011) Dendritic coding of multiple sensory inputs in single cortical neurons in vivo. Proc Natl Acad Sci USA 108:15420–15425

Wachowiak M, Denk W, Friedrich RW (2004) Functional organization of sensory input to the olfactory bulb glomerulus analyzed by two-photon calcium imaging. Proc Natl Acad Sci USA 101:9097–9102

Chapter 6
Optical Imaging Techniques for Investigating the Function of Earthworm Nervous System

Kotaro Oka and Hiroto Ogawa

Abstract The earthworm is a suitable animal for investigating the relationships between its behavior and the function of its nervous system. However, this animal has not been widely used for neuroethological research because it is not a model organism. In this chapter, we describe the use of the earthworm as a plausible model for investigating locomotion activity and learning and memory abilities using several optical imaging techniques based on our previous studies. First, we demonstrate the usefulness of this animal briefly, show its simple nervous system, and then focus on its locomotion mechanism in detail. Especially, we focus on experimental techniques using several fluorescent dyes, e.g., FM1-43 for functional imaging of specific synapses and DAF for nitric oxide release from the ventral nervous system following different forms of stimulation. Finally, we conclude this chapter with a discussion on future studies of this animal with respect to learning and memory and imaging techniques.

Keywords Activity-dependent labeling • Ca imaging • Earthworm • FM dyes • Lipophilic fluorescent dye • Nitric Oxide (NO)

6.1 Introduction

One of the first and most famous experimenters using the earthworm was Charles Darwin. He investigated the earthworm for approximately 40 years after he published a book on soil in 1837. He published his famous book *The Formation of*

K. Oka (✉)
School of Fundamental Sciences and Technology, Faculty of Science and Technology,
Keio University, 3-14-1 Hiyoshi, Kohoku-ku, Yokohama 223-8522, Japan
e-mail: oka@bio.keio.ac.jp

H. Ogawa
Department of Biological Science, Faculty of Science, Hokkaido University,
Kita 10, Nishi 8, Kita-ku, Sapporo 060-0810, Japan

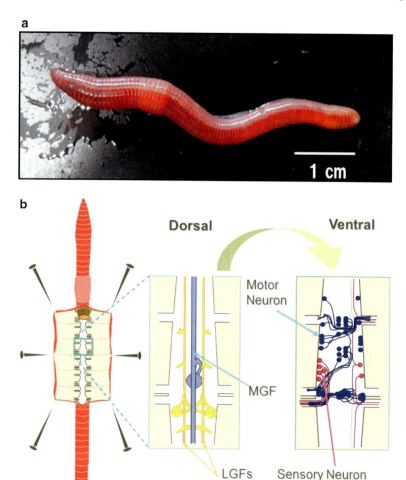

Fig. 6.1 Earthworm, *Eisenia fetida* (**a**), and its nervous system (**b**)

Vegetable Mould through the Action of Worms, with Observations on their Habits in 1881 as his last study. In this book, he mentioned the intelligence of this small animal, partly concerning their learning and memory abilities.

On the other hand, many people have their first experience performing biological experiments using the earthworm because of the frequent use of this animal as a specimen for physiological experiments during undergraduate studies. The earthworm has three giant nerve fibers: one medial giant fiber (MGF) and a pair of lateral giant fibers (LGFs, Fig. 6.1) (Edwards and Loftly 1972; Mill 1982). We can measure the propagation of action potentials along these giant fibers using a simple nerve box as a source of extracellular potential, and this simple experimental setup has been used for elementary experiments for newcomers to the Department of Biology.

However, the neuronal function of the earthworm and its relationship to behavior are still mostly unknown.

Therefore, we have investigated the relationship between the nervous system and behavior of the earthworm, *Eisenia fetida*, for the last 20 years using conventional electrophysiology and optical imaging techniques for Ca, nitric monoxide (NO), immunohistochemistry, and activity-dependent presynaptic imaging. In this chapter, we describe the strategies we have used to understand the behavior of this animal with several powerful imaging techniques.

6.2 Optical Imaging of the Nervous System Using Specific Fluorescent Dyes

Figure 6.2 contains a summary of several imaging techniques that are used to investigate neuronal networks and their functions. Electron microscopy is the basic tool for investigating synaptic connections in the nervous system (Fig. 6.2a). This figure reveals the huge axons of the MGF and LGFs in the ventral nerve cord (VNC) and the complicated structure of the neuropile. Because electron microscopy is the only method that can be used to identify the location of synapses and to understand, the reconstruction of the synaptic connections is a laborious task. Recently, the development of low-pressure scanning electron microscopy and sequential large-field sectioning instruments has enabled the high-throughput reconstruction of the nervous system, and this approach has been named as "connemics" or the "connectome" (Lichtman and Denk 2011).

For neuroethological investigations, knowledge of the innervations and projections between neurons is crucial. One approach to visualize the nervous system is the application of several kinds of fluorescent dyes to the target neurons. Figure 6.2b illustrates the application of two types of lipophilic fluorescent dyes (DiI and DiA) to the cut ends of the left and right segmental nerves. This double-staining technique can visualize projection neurons to each and also both sides of the muscles and body wall. The cell bodies of these projection neurons are clearly visualized in vivo, and the combination of this technique with intracellular recording enables the investigation of the neural network and its function simultaneously. Furthermore, by backfilling with fluorescent Ca indicators from a specific segmental nerve, we can investigate neuronal activity and also functional synaptic connections that are related to specific earthworm behavior.

Without staining, we can visualize neurons and several fine axons in the nervous system. The nervous system of the earthworm is opaque, and the insertion of an intracellular electrode or application of a patch electrode is difficult without staining. However, by using differential interference contrast (DIC) microscope with infrared illumination, we can easily apply electrophysiological techniques to the nervous system without staining (Fig. 6.2c). After the electrophysiological experiment is complete, microinjection of a fluorescent dye from the microelectrode enables the visualization of the structure of the target neuron in detail.

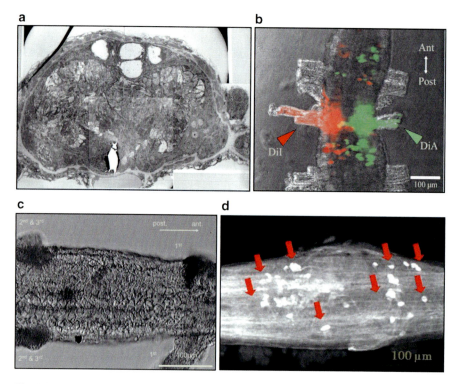

Fig. 6.2 Several types of imaging techniques that are useful for neuroethological studies. (**a**) Cross section of the ventral nerve cord (VNC) of the earthworm. MGF and a pair of lateral giant fibers (LGFs) are observed on the dorsal side. Cell bodies of neurons with huge nuclei are located in the peripheral region of the VNC. (**b**) Retrograde trace of neurons using lipophilic fluorescent dyes (DiI and DiA). Two different dyes were applied to the right and left cut ends of the first segmental nerves, and the cell bodies of the projection neurons are imaged with *red* and *green colors*. In the middle part of the VNC, a yellow-colored neuron indicates a projection neuron to both the right and left segmental nerves. (**c**) Differential interference contrast microscopy with infrared illumination. This method is useful for inserting a microelectrode into specific neurons without staining. (**d**) Immunohistochemistry of the VNC using an anti-serotonin antibody. This old method has been revived following the development of several new fluorophores for multicolor staining

Finally, we do not forget the use of conventional immunohistochemistry (Fig. 6.2d). In this figure, neurons containing the neurotransmitter serotonin (5-HT) are visualized. Although this is an old technique, a combination of this technique and other imaging and/or electrophysiology methods is fundamental for neuroethological investigations characterizing specific neurons. In the ventral nervous system of the earthworm, we found specific neurons with two neurotransmitters, 5-HT and FMRF amide, that control the circular and longitudinal muscles, respectively. These specific neurons are important for the coordinated contraction of muscles during locomotion.

6.3 NO Release from the Nervous System

In this section, we present an imaging technique for visualizing neuronal function, especially signal transduction. Several types of fluorescent indicators have been developed and applied in neuroethological studies. Ca indicators have been used to visualize multi-neuron activity in the nervous system, and some of their applications are presented in another chapter of this book. We, therefore, present other kinds of fluorescent dyes in this chapter.

NO is responsible for various types of neuromodulation in vertebrate and invertebrate nervous systems (for reviews, Garthwaite and Boulton 1995; Jaffrey and Snyder 1995). NO was also found to be involved in the chemical modification of serotonin into inactivate forms (Blanchard et al. 1997). For example, the neuromodulatory effect of serotonin on cholinergic synapses in the buccal ganglion of *Aplysia* is reduced in the presence of an NO donor due to serotonin inactivation by NO-derived nitrogen oxide (Fossier et al. 1999). In the distal colon of the guinea pig, the secretion of chloride by serotonin is possibly induced via the activation of NO-producing neurons (Kadowaki et al. 1996; Kuwahara et al. 1998). Evidence of a physiological interaction between NO and serotonin is also supported by the morphological relationships between NO synthase -containing neurons and serotonergic neurons (Kadowaki et al. 1999).

At the present time, there are two plausible methods to detect NO: NO-sensitive dye staining and NO-selective microelectrodes. The first method is convenient for screening the production of NO, and the second method is useful for the quantitative evaluation of NO release from the nervous system.

We successfully showed that the VNC of the earthworm produces NO as a neuromodulator by visualizing the spatial pattern of NO production in the VNC using an NO-specific fluorescent dye, DAF-2 DA (Kitamura et al. 2001a, b). The mechanism for detecting NO with DAF-2 DA and typical imaging of spontaneous NO production in the ventral nerve cord (VNC) of the earthworm are illustrated in Fig. 6.3a, b, respectively. This previous study revealed that the VNC contains approximately 70 nitrergic neurons on the ventral side, which were identified using nicotinamide adenine dinucleotide phosphate diaphorase (NADPH-d) histochemistry, and suggested that basal NO production from these neurons is relatively high compared with other previously reported nervous systems (Moroz et al. 1995; Kobayashi et al. 2000a, b). It also showed that high K^+ stimulation induces further NO production, resulting in the diffusion of NO to the giant fibers on the dorsal side. The axons of the ventrolateral serotonergic cells in the VNC are connected to the MGF (Lubics et al. 1997), and the propagation of action potentials in the MGF can be modulated by the addition of 10 μM serotonin. These findings suggest a physiological interaction between NO and serotonin in the earthworm VNC. However, the role of serotonin in the mechanism of NO production in nervous systems, including that of the earthworm VNC, has not been investigated. Recently, we found that inhibition of serotonin-induced NO production causes the suppression of

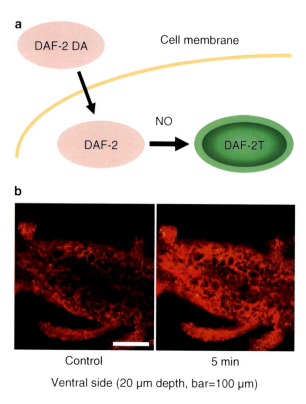

Fig. 6.3 Nitric oxide (NO) detection using a fluorescent probe. (**a**) Mechanism for detecting the release of NO using the fluorescent probe DAF-2. DAF-2 DA is cell membrane permeable because of its acetoxymethyl residue. After crossing the cell membrane, DAF-2 DA is transformed into DAF-2, which is not membrane permeable, by esterase activity in the cell. Finally, DAF-2 captures NO and changes its form to DAF-2T and fluoresces. Notice that the fluorescent intensity of DAF-2T does not decrease with NO decrease. (**b**) Typical imaging of spontaneous NO production in the ventral nerve cord of the earthworm (modified from Kitamura et al. 2001a, b)

associative learning between vibration (conditioned stimulus) and light (unconditioned stimulus) in the earthworm. Furthermore, the injection of serotonin into the body cavity enhances the acquisition of this associative learning. These results indicate the essential interest in NO and associative learning, and quantitative investigations between NO production and learning are required. We, therefore, developed an NO-selective microelectrode (Kitamura et al. 2000, 2003) for further study.

6.4 Activity-Dependent Observation of the Presynaptic Region with FM1-43

The typical locomotory behavior of the earthworms is creeping. The earthworm shrinks and elongates its body wall periodically with the regular contraction of the circular and longitudinal muscles. We found that the neural activity of this muscle contraction is induced by a neuromodulator, octopamine, without effectors or sensory feedback (fictive locomotion). The application of octopamine to the isolated VNC induces rhythmic neuronal activity observed in the segmental nerve cord, and the frequency of this neuronal activity depends on the concentration of octopamine (Mizutani et al. 2002,

6 Optical Imaging of Earthworm Nervous System 95

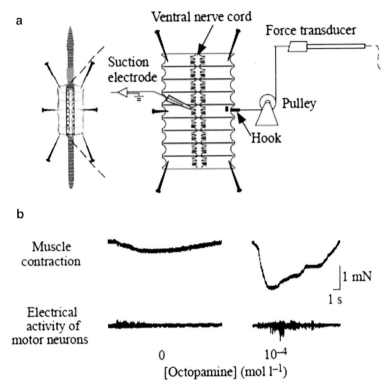

Fig. 6.4 Circular muscle contraction induced by octopamine. (**a**) Diagram of a semi-intact preparation. A small thread which is connected to a force transducer via a pulley is hooked in the middle part of the dissected body wall. The electrical activity of motor neurons was recorded simultaneously from the cut end of the first segmental nerve in the same segment using a glass suction electrode. (**b**) The simultaneous recording of muscle contraction and electrical activity from projecting nerves. In control preparations, electrical activity was only seen at the onset of a contraction. At 10^{-4} M octopamine, electrical activity occurred throughout much of the contraction period (modified from Mizutani et al. 2002)

2004). The simultaneous measurement of muscle contraction and neuronal activity from the segmental nerve cord is shown in Fig. 6.4a. Previously, we confirmed that the neuronal activity from the left and right segmental nerve cord is coincident during fictive locomotion, and octopamine induces the huge contraction of the circular muscles with more frequent neuronal activity (Fig. 6.4b). However, the motor neurons and interneurons underlying fictive locomotion were still unknown at that time; therefore, we tried to identify such neurons using optical imaging techniques.

The development of various imaging techniques and fluorescent dyes has enabled researchers to visualize the mobilization of various substances in neurons. We have various choices for Ca indicators, for example, Fura-2, Fluo-3 and Fluo-4, and Calcium Green-1, and also several kinds of fluorescent protein-based Ca sensors, including Cameleon (Miyawaki et al. 1997). We previously examined Ca wave propagation induced by electrical tetanic stimulation in the earthworm giant axon using Calcium Green-1 (Ogawa et al. 1994; Ogawa and Oka 1996). Furthermore, by using voltage-dependent fluorescent dyes (RH and JPW series dyes, see recent review, Chemla and Chavane 2010) and voltage-sensitive fluorescent proteins (VSFP series proteins, see Mutoh et al. 2011), direct measurement of membrane potential as an indicator of neuronal activity has been developed. Optical recording using voltage-sensitive dyes and proteins enables us to study the spatial pattern of stimuli-induced depolarization and hyperpolarization at a high time resolution. However, the location of individual neurons activated by the mechanical stimuli and their connectivity remains unknown. Lucifer yellow and sulforhodamine are often used as markers of neuronal activity (Wilcox and Franceschini 1984; Keifer et al. 1992; Kimura et al. 1998). On the other hand, the styryl dye N-(3-triethyl-ammoniumpropyl)-4-(4-(dibutylamino)styryl)-pyridinium dibromide (FM1-43, Fig. 6.5a is useful for visualizing active synapses (Betz and Bewick 1992; Betz et al. 1992). When the internalization of membrane vesicles occurs in the presence of FM1-43, the dye is trapped inside the endocytosed vesicles. Then, when the dye is removed from the external medium, the FM1-43-labeled vesicles can be detected fluorescently (Betz and Bewick 1992; Betz et al. 1992). The fluorescence intensity decreases when the dye-loaded vesicles are exocytosed again and release the dye to the outside medium (Betz and Bewick 1992; Betz et al. 1992; Klingauf et al. 1998). We applied FM1-43 staining to the VNC of the earthworm and confirmed the neuronal activity-dependent staining of the VNC with FM1-43 (Shimizu et al. 1999). The same technique was also applied to intact neural preparations from *Caenorhabditis elegans*, lamprey, and rat (Kay et al. 1999). We, therefore, applied activity-dependent labeling with FM1-43 to the nervous system of the earthworm to determine the location of neural activity during fictive locomotion.

We investigated fluorescent staining of the VNC with FM1-43 and observed the fluorescent spots as a function of the concentration of octopamine (Mizutani et al. 2003). The intensity of each fluorescent spot increased with increasing octopamine concentration. An analysis of the fluorescence intensity of the spots relative to octopamine concentration showed that two response patterns were present (Fig. 6.5b). One group of spots responded at low concentrations of octopamine (10^{-7} M) and was saturated at 10^{-5} to 10^{-4} M (high-affinity group; Fig. 6.5b, plotted as diamonds). The other group of spots responded from 10^{-5} M and was saturated at 10^{-4} to 10^{-3} M (low-affinity group; plotted as triangles), in a manner similar to the sensitivity of the motor pattern response to exposure to octopamine (plotted as circles). The octopamine concentrations for the half-maximal responses in the high- and low-affinity groups were 1.0×10^{-6} M and 2.0×10^{-5} M, respectively, while that of the motor pattern was 1.0×10^{-6} M.

The voltage-sensitive dye ww-781 can be applied after monitoring FM1-43 uptake to visualize the location of neurons and presynaptic regions. Because ww-781 is lipophilic and stains neuronal cell membranes, we inverted the acquired fluorescence images to visualize neuronal cell bodies and then superimposed the FM1-43 fluorescent images over the ww-781 inverted fluorescent images (Fig. 6.5c). FM1-43 spots from the group with a high affinity to octopamine were located at 10–15 μm from the edge of the VNC, between the first and second/third lateral nerves (Fig. 6.5c, yellow area), while those from the low-affinity group were located

6 Optical Imaging of Earthworm Nervous System

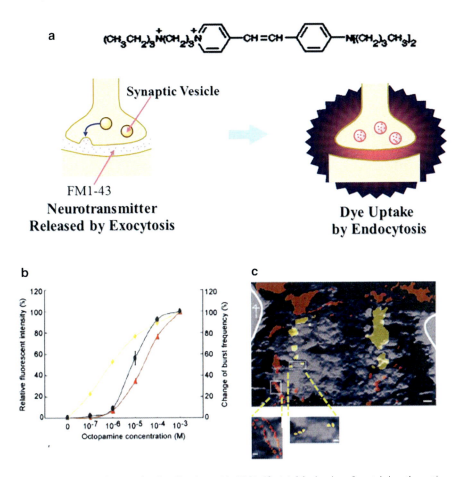

Fig. 6.5 Functional synaptic visualization with FM1-43. (**a**) Mechanism for staining the active presynaptic terminal with FM1-43. (**b**) Dose dependence of the relative fluorescence intensity as a function of octopamine concentration. Two-response patterns were observed. One group of spots (high affinity; *diamonds*) responds to low concentrations of octopamine and saturates at 10^{-5} to 10^{-4} M. The other group (low affinity; *triangles*) responds to octopamine concentrations in excess of 10^{-5} M and saturates at 10^{-4} to 10^{-3} M, in a manner similar to the motor pattern response (*circles*). The *bar* indicates S.D. (**c**) The *yellow areas* indicate the FM1-43 fluorescent spots with a high affinity to octopamine. The *red areas* indicate low-affinity regions. The *blue background* is the inverted ww-781 fluorescence image. Scale bar = 5 µm. The *arrowhead* indicates the anterior side (modified from Mizutani et al. (2003))

at 5–10 µm from the edge of the VNC, between the first and second/third lateral nerves and the center of the VNC (Fig. 6.5c, red area). We magnified the red and yellow fluorescent spots in the rectangular areas of Fig. 6.5c by a factor of three and show the area with a higher fluorescence intensity in Fig. 6.5c. In this way, bright fluorescent FM1-43 spots of 1–2 µm in diameter were observed on the cell membrane of neuronal cell bodies.

6.5 Future Directions of Earthworm Neuroethology with Imaging Techniques

In this chapter, we briefly introduced several imaging techniques that can be used to understand the function of the ventral nervous system of the earthworm. These experimental techniques are fundamental and nonrestrictive for studies of the earthworm, and combinatorial approaches with these methods will become powerful tools to understand the behavior of this animal. For example, behavioral experiments with FM1-43 staining may reveal the functional networks for specific behaviors. According to this line of work, we aim to visualize the specific neural networks for fictive locomotion and also for associative learning between vibration and light. Furthermore, Ca imaging is useful to observe the neural activity from many neurons simultaneously. We successfully visualized several neurons using a dextran-conjugated Ca indicator that was applied to the cut end of the segmental nerve cord and measured neuronal activity during fictive locomotion. The combination of conventional electrophysiology, Ca imaging, and FM1-43 will clarify the neuronal basis of the behavior of the earthworm.

References

Betz WJ, Mao F, Bewick GS (1992) Activity-dependent fluorescent staining and destaining of living vertebrate motor nerve terminals. J Neurosci 12:363–375

Betz WJ, Bewick GS (1992) Optical analysis of synaptic vesicle recycling at the frog neuromuscular junction. Science 255:200–203

Blanchard B, Dendane M, Gallard JF, Houee-Levin C, Karim A, Payen D, Launay JM, Ducrocq C (1997) Oxidation, nitrosation, and nitration of serotonin by nitric oxide-derived nitrogen oxides: biological implications in the rat vascular system. Nitric Oxide 1:442–452

Chemla S, Chavane F (2010) Voltage-sensitive dye imaging: technique review and models. J Physiol Paris 104:40–50

Edwards CA, Loftly JR (1972) Biology of earthworms. Chapman & Hall, London

Fossier P, Blanchard B, Ducrocq C, Leprince C, Tauc L, Baux G (1999) Nitric oxide transforms serotonin into an inactive form and this affects neuromodulation. Neuroscience 93:597–603

Garthwaite J, Boulton CL (1995) Nitric oxide signaling in the central nervous system. Annu Rev Physiol 57:683–706

Jaffrey SR, Snyder SH (1995) Nitric oxide: a neural messenger. Annu Rev Cell Dev Biol 11:417–440

Kadowaki M, Gershon MD, Kuwahara A (1996) Is nitric oxide involved in 5-HT-induced fluid secretion in the gut? Behav Brain Res 73:293–296

Kadowaki M, Kuramoto H, Kuwahara A (1999) Morphological relationship between serotonergic neurons and nitrergic neurons for electrolytes secretion in the submucous plexus of the guinea pig distal colon. Brain Res 831:288–291

Kay AR, Alfonso A, Alford S, Cline HT, Holgado AM, Sakmann B, Snitsarev VA, Stricker TP, Takahashi M, Wu LG (1999) Imaging synaptic activity in intact brain and slices with FM1-43 in C. elegans, lamprey, and rat. Neuron 24:809–817

Keifer J, Vyas D, Houk JC (1992) Sulforhodamine labeling of neural circuits engaged in motor pattern generation in the in vitro turtle brainstem-cerebellum. J Neurosci 12:3187–3199

Kimura T, Suzuki H, Kono E, Sekiguchi T (1998) Mapping of interneurons that contribute to food aversive conditioning in the slug brain. Learn Mem 4:376–388

Kitamura Y, Uzawa T, Oka K, Komai Y, Ogawa H, Takizawa N, Kobayashi H, Tanishita K (2000) Microcoaxial electrode for in vivo nitric oxide measurement. Anal Chem 72:2957–2962

Kitamura Y, Naganoma Y, Horita H, Tsuji N, Shimizu R, Ogawa H, Oka K (2001a) Visualization of nitric oxide production in the earthworm ventral nerve cord. Neurosci Res 40:175–181

Kitamura Y, Naganoma Y, Horita H, Ogawa H, Oka K (2001b) Serotonin-induced nitric oxide production in the ventral nerve cord of the earthworm, *Eisenia fetida*. Neurosci Res 41:129–134

Kitamura Y, Ogawa H, Oka K (2003) Real-time measurement of nitric oxide using a bio-imaging and an electrochemical technique. Talanta 61:717–724

Klingauf J, Kavalali ET, Tsien RW (1998) Kinetics and regulation of fast endocytosis at hippocampal synapses. Nature 396:581–585

Kobayashi S, Ogawa H, Fujito Y, Ito E (2000a) Nitric oxide suppresses fictive feeding response in *Lymnaea stagnalis*. Neurosci Lett 285:209–212

Kobayashi S, Sadamoto H, Ogawa H, Kitamura Y, Oka K, Tanishita K, Ito E (2000b) Nitric oxide generation around buccal ganglia accompanying feeding behavior in the pond snail, *Lymnaea stagnalis*. Neurosci Res 38:27–34

Kuwahara A, Kuramoto H, Kadowaki M (1998) 5-HT activates nitric oxide-generating neurons to stimulate chloride secretion in guinea pig distal colon. Am J Physiol 275:G829–G834

Lichtman JW, Denk W (2011) The big and the small: challenges of imaging the brain's circuits. Science 334:618–623. doi:10.1126/science.1209168

Lubics A, Reglodi D, Slezak S, Szelier M, Lengvari I (1997) Co-localization of serotonin and FMRFamide-like immunoreactivities in the central nervous system of the earthworm, *Eisenia fetida*. Acta Histochem 99:459–467

Mill PJ (1982) Recent developments in earthworm neurobiology. Comp Biochem Physiol 73A:641–661

Miyawaki A, Llopis J, Heim R, McCaffery JM, Adams JA, Ikura M, Tsien RY (1997) Fluorescent indicators for Ca^{2+} based on green fluorescent proteins and calmodulin. Nature 388:882–887

Mizutani K, Ogawa H, Saito J, Oka K (2002) Fictive locomotion induced by octopamine in the earthworm. J Exp Biol 205:265–271

Mizutani K, Shimoi T, Kitamura Y, Ogawa H, Oka K (2003) Identification of two types of synaptic activity in the earthworm nervous system during locomotion. Neuroscience 121:473–478

Mizutani K, Shimoi T, Kitamura Y, Ogawa H, Oka K (2004) Modulation of motor patterns by sensory feedback during earthworm locomotion. Neurosci Res 48:457–462

Moroz LL, Radbourne S, Winlow W (1995) The use of NO-sensitive microelectrodes for direct detection of nitric oxide (NO) production in molluscs. Acta Biol Hung 46:155–167

Mutoh H, Perron A, Akemann W, Iwamoto Y, Knöpfel T (2011) Optogenetic monitoring of membrane potentials. Exp Physiol 96:13–18

Ogawa H, Oka K, Fujita S (1994) Calcium wave propagation in the axon of the earthworm. Neurosci Lett 179:45–49

Ogawa H, Oka K (1996) Physiological basis of earthworm ventral nerve cord I. Heterogeneity of cellular mechanisms underlying calcium mobilization in the median giant fiber. Bioimages 4:137–147

Shimizu R, Oka K, Ogawa H, Suzuki K, Saito J, Mizutani K, Tanishita K (1999) Optical monitoring of the neural activity evoked by mechanical stimulation in the earthworm nervous system with a fluorescent dye, FM1-43. Neurosci Lett 268:159–162

Wilcox M, Franceschini N (1984) Illumination induces dye incorporation in photoreceptor cells. Science 225:851–854

Part IV
Optogenetics

Chapter 7
Monitoring Neural Activity with Genetically Encoded Ca^{2+} Indicators

Azusa Kamikouchi and André Fiala

Abstract Visualizing the activity of nerve cells using genetically encoded indicator proteins has emerged to a widely used technique in the field of neuroscience. In particular, intracellular Ca^{2+} dynamics represents a parameter that is closely correlated with neuronal excitation, and a variety of genetically encoded Ca^{2+} sensors have been developed. The fruit fly *Drosophila melanogaster* is an extremely useful model organism to use these indicators because of its sophisticated genetic tools to express an artificial genetic construct in a spatially and temporally controlled pattern within the nervous system. Binary expression systems, for which large amount of different fly strains exist, enable a targeted expression in selective neuronal populations. Advanced fluorescence microscopical visualization techniques (see Part 3) allow for real-time monitoring of neural activity patterns. In *Drosophila*, optical Ca^{2+} imaging has been used to analyze basic principles of neuronal coding and processing, e.g., olfactory coding, visual stimulus processing, taste perception, mechanosensation, or learning and memory. In this chapter, we will review how genetic targeting methods can be used in *Drosophila* to monitor neural Ca^{2+} activity in vivo in order to study how individual neurons or neuronal ensembles encode stimulus information.

A. Kamikouchi (✉)
Division of Biological Science, Graduate School of Science, Nagoya University, Chikusa, Furo, Aichi 464-8602, Japan

Precursory Research for Embryonic Science and Technology, Japan Science and Technology Agency, Tokyo 102-0076, Japan
e-mail: kamikouchi@bio.nagoya-u.ac.jp

A. Fiala
Department of Molecular Neurobiology of Behavior, Georg-August University, Schwann-Schleiden Forschungszentrum, Julia-Lermontowa-Weg 3, 37077 Göttingen, Germany

Keywords Drosophila • GAL4/UAS system • Genetically encoded Ca^{2+} indicator (GECI) • Imaging

7.1 Introduction

Drosophila melanogaster is one of the most important model organisms for various fields of biological research, such as genetics, developmental biology, or neuroscience. The fruit fly is easy to breed, and it produces much offspring in short generation cycles. Under ideal conditions, the development from egg to adult takes only ca. 10 days at 25 °C. Moreover, the fruit fly exhibits a wide array of stereotyped and nonstereotyped, flexible behaviors, covering simple reflexes such as phototaxis (Benzer 1967) or chemotaxis (Vosshall and Stocker 2007); circadian behavior and sleep (Hendricks et al. 2000); more complex behavioral patterns, e.g., courtship (von Philipsborn et al. 2011) or aggression (Dankert et al. 2009; Hoyer et al. 2008); and behavioral plasticity, e.g., olfactory learning and memory formation (Fiala 2007).

The brain of the fruit fly, which controls these behaviors, is composed of ca. 100,000 neurons (Chiang et al. 2011; Shimada et al. 2005). As a general principle, their cell bodies are distributed at the outer surface of the brain, whereas their processes innervate the interior side of the brain and form the synaptic neuropils. Decades of research have led to a fairly good description of the neuronal structures and connectivities within the *Drosophila* brain. A recent construction of a mesoscopic map of the fruit fly brain has, for example, suggested that the *Drosophila* brain consists of 41 local processing units, six hubs, and 58 tracts (Chiang et al. 2011). It is obvious that the numbers of nerve cells and structurally identifiable brain regions are much smaller than those of higher vertebrates, which facilitates of course the analysis of circuits underlying behavior. However, the most important advantage of *Drosophila* when compared with other organisms relies on sophisticated gene-expression systems that can be used, for example, to monitor and manipulate neural activity in vivo.

Genetic germline transformation techniques based on the *P* element transposon (Spradling and Rubin 1982) have initially been used to identify and disrupt genes and their regulatory elements. But *P* elements carrying regulatory elements are also used to induce gene expression, e.g., of fluorescent sensor proteins, in a cell type-specific spatiotemporal pattern (Duffy 2002; Venken et al. 2011). The combination of a wide array of behavioral paradigms, a relatively small brain, and sophisticated expression systems with new tools to monitor and manipulate neural activity turns *Drosophila* into an excellent model system to study the complexity of a brain in detail. This chapter will provide a brief introduction in *Drosophila* genetics, a summary of the available genetic tools to visualize neural activity, and examples of their applications in *Drosophila* neurobiology.

7.2 Gene-Expression Systems in *Drosophila*

Specific behaviors of animals are often generated and organized by activity patterns across ensembles of neurons mediating sensory processing, decision-making, and motor control. Analyzing the neural processes that underlie a specific behavior thus requires measuring neural activity in vivo. In *D. melanogaster*, transgenes for monitoring neural activity can be targeted to almost any population of neurons using binary expression systems (Venken et al. 2011).

A binary expression system consists of a transcription activator and its target sequence to which the transcription activator binds. These two elements are separated on two different transgenic *Drosophila* strains, a "driver strain" that determines where and when the gene of interest is expressed and a "responder strain" that determines which gene of interest is actually expressed. The GAL4/UAS system was the first and is still the most widely used binary system developed for *D. melanogaster* (Brand and Perrimon 1993) (Fig. 7.1). The key for this was the finding that the transcription activator GAL4 identified in the yeast *Saccharomyces cerevisiae* functions also well in the fruit fly when a responsive DNA element for it, called the upstream activation sequence (UAS) element, is present (Fischer et al. 1988). In the GAL4/UAS system, the two elements are now separated on a GAL4 strain ("driver strain") and a UAS strain ("responder strain"). If both strains are crossed, the F1 generation carries both transgenic DNA insertions, and the transcription factor GAL4 can bind to the UAS sequence that induces the expression of the desired gene, e.g., a Ca^{2+} sensor (Fig. 7.1a, b).

This principle is not limited to *Drosophila,* and the GAL4/UAS system is also used in other genetically tractable organisms such as zebrafish or mice (Ornitz et al. 1991; Scheer and Campos-Ortega 1999). The beauty of the GAL4/UAS system in *Drosophila*, however, relies on a very large collection of fly strains; thousands of GAL4 strains have been created by random *P* element insertion, and each strain expresses GAL4 in a designated spatiotemporal pattern (Hayashi et al. 2002) (Fig. 7.1c). Moreover, a variety of defined promoters, such as the pan-neuronal ELAV promoter (Yao and White 1994) and the eye-selective GMR promoter (Hay et al. 1994), have been used to generate tissue-specific GAL4 strains. Recently, GAL4 strains that drive gene expression in more restricted patterns were systematically generated by inserting small fragments from the flanking noncoding and intronic regions of genes (Pfeiffer et al. 2008). This can lead to a transgene expression in small subsets of cells, e.g., very few selected neurons within the brain. Furthermore, the efficacy of the GAL4 system was improved by optimizing factors that affect the pattern and strength of GAL4-driven expression such as codon usage, mRNA stabilization, transcription activation domain, and the number of UAS sites (Pfeiffer et al. 2010). A variety of additional techniques have been developed to further narrow down the expression of the transgene of interest in time and space. Genetic mosaics can be created using the MARCM system (Wu and Luo 2006) that

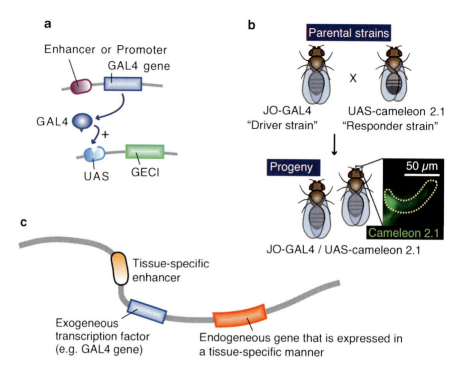

Fig. 7.1 Genetic tools to express a GECI construct in tissue-specific manner. (**a**) Illustration of the binary GAL4/UAS expression system. In the transgenic progeny obtained from a genetic cross of a GAL4 strain and a UAS-GECI strain, the indicator is expressed in a specific cell population defined by the parental GAL4 strain (Brand and Perrimon 1993). (**b**) A genetic cross to obtain Cameleon 2.1 expression in Johnston's organ (JO) neurons. JO-GAL4 is a driver strain that expresses GAL4 in JO neurons (Kamikouchi et al. 2009). When this GAL4 fly is crossed to the responder strain UAS-Cameleon 2.1 (Fiala and Spall 2003), progeny containing both elements is produced. The presence of GAL4 drives expression of Cameleon 2.1 (*green*) in the JO neurons. The *dotted line* indicates the array of cell bodies of JO neurons. (**c**) The enhancer-trapping method. A P element carrying an exogenous transcription factor such as the GAL4 transcriptional activator is randomly mobilized throughout the genome, bringing the expression of GAL4 under the control of endogenous tissue-specific enhancers (Duffy 2002)

enables a transgene expression in clones of neurons derived from a common ancestor cell (Kohatsu et al. 2011). Various so-called intersectional strategies have also been developed to refine the pattern of GAL4-driven gene expression, such as the expression of the GAL4 inhibitor GAL80 in addition to GAL4 but in a different, overlapping subset of cells (Lee and Luo 1999), or the random induction of transgene expression using a flippase (Flp) and other recombinases (Nern et al. 2011). In the "split GAL4" system, a functional GAL4 protein can also be created by an independent expression of two parts of it, and the expression of both parts can be genetically targeted to different, overlapping subsets of cells (Luan et al. 2006). Lastly, alternative binary systems have recently been described, such as the LexA system and the QF system (Lai and Lee 2006; Potter et al. 2010). All of those technical refinements allow for a very precisely controllable expression of the transgene of interest, which can, of course, be a Ca^{2+} sensor protein.

7.3 Genetically Encoded Ca^{2+} Indicators

Genetically encoded Ca^{2+} indicators (GECIs) [also called fluorescent calcium indicator proteins (FCIPs)] are widely used to image activity in defined neuronal populations. GECIs consist of a calcium-binding domain such as calmodulin or troponin C, which is fused to GFP-derived fluorescent proteins (Mank et al. 2008; Miyawaki et al. 1997; Nakai et al. 2001). Two principal kinds of Ca^{2+} sensor proteins have been described, single-chromophore sensors and two-chromophore sensors. In single-fluorescent-protein indicators, the fluorescence intensity of a modified fluorescence protein is modulated by calcium-binding-dependent changes in the chromophore environment. These indicators show very low fluorescence at the unbound state but increase their emission intensities drastically after binding calcium. Among them, the GCaMP (Nakai et al. 2001) is the most widely used one across multiple model organisms from nematodes, fruit flies, zebrafish to mice (Hasan et al. 2004; Higashijima et al. 2003; Kerr et al. 2000; Wang et al. 2003). GCaMP consists of an enhanced GFP that has been circularly permuted (cpEGFP) such that new N- and C-termini have been introduced. To the new termini, a calmodulin sequence and the calmodulin-binding M13 fragment of myosin light chain kinase have been fused (Nakai et al. 2001). When Ca^{2+} binds to calmodulin, conformational changes due to the Ca^{2+}–calmodulin–M13 interaction induce a subsequent conformational change in cpEGFP such that the emission intensity changes when the chromophore is excited. Site-directed mutagenesis has improved the GCaMP indicators significantly (Muto et al. 2011; Tian et al. 2009). In *Drosophila*, the version GCaMP3 is widely used to monitor neural Ca^{2+} activity in intact flies (Chiappe et al. 2010; Seelig et al. 2010), and it shows high signal-to-noise ratio, high sensitivity with respect to transient neural activity, and fast kinetics in vivo (Tian et al. 2009).

In two-chromophore indicators, calcium binding modulates the efficiency of Förster resonance energy transfer (FRET) between a pair of fluorescent proteins, usually an enhanced cyan (eCFP) and an enhanced yellow (eYFP) fluorescent protein (Miyawaki et al. 1999; Miyawaki et al. 1997), between which a calcium-binding domain is sandwiched. If the eCFP (donor chromophore) is excited at ~440 nm wavelength, the emission light energy is transferred to the eYFP (acceptor chromophore) in the form of FRET, which decreases the emission intensity from the donor chromophore and increases the emission intensity from the acceptor chromophore. As a result the ratio of emissions between these chromophores reflects the extent by which intracellular calcium concentration changes. Such ratiometric, FRET-based Ca^{2+} indicators have potential advantages over single-fluorescent-protein GECIs, including higher baseline brightness and relative insensitivity to motion artifacts

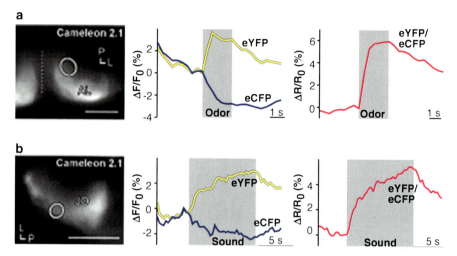

Fig. 7.2 Stimulus-evoked calcium signals. Ca^{2+} signals in the AL (**a**) and the JO neurons (**b**) in response to odor and sound, respectively, were recorded. The *circles* in the *left panels* indicate the regions in which calcium signals were monitored. *Dashed line* indicates the midline of the brain. Each stimulus-evoked reciprocal fluorescent changes ($\Delta F/F_0$) between eCFP (*blue line*) and eYFP (*yellow line*) by FRET (*middle panels*). $\Delta R/R_0$ (%) is the change in eYFP/eCFP fluorescence ratio, where R is the average eYFP/eCFP ratio before stimulus onset and ΔR is the deviation from R (*right panels*). Scale bar in the *left panels* = 50 μm. Lines with characters L and P in the left panels indicate the directions to the lateral and posterior, respectively [modified from Kamikouchi and Ito (2007)]

because of wavelength ratioing. A drawback of the ratiometric indicators relies on the relatively small fluorescence changes caused by FRET, which results in a lower signal intensity when compared to single-chromophore sensors. Calmodulin-based two-chromophore indicators, such as Cameleon (Fiala et al. 2002; Miyawaki et al. 1999) and D3cpVenus (Hendel et al. 2008; Palmer et al. 2006), or the troponin C-based indicator such as TN-XXL (Mank et al. 2008) has been successfully used in *Drosophila*. Technically, two-chromophore sensors are more complicated to use as its application requires the simultaneous recording of two wavelengths. A CCD camera with a beam splitter device or two detectors with different emission filters enables such dual-channel imaging.

By choosing GAL4 fly strains from the available collections, GECIs can be targeted to virtually any neurons of interest (Fig. 7.2). These probes translate a change in ion concentration into changes in the fluorescence (Palmer and Tsien 2006). Thereby, the activity of large populations of genetically and functionally related neurons can be monitored simultaneously, which overcomes the spatial limitations of electrode recordings. Whereas electrophysiological recordings using electrodes monitor changes in membrane or field potential of individual neurons or neuronal populations, Ca^{2+} sensors exploit the fact that changes in membrane potential are typically accompanied by an increase in intracellular Ca^{2+} concentrations. Moreover, genetic probes are continuously expressed in vivo, which facilitates long-term and repeated measurements in genetically specified neurons. On the other hand, the main advantage of electrode recordings—the enormously high sensitivity and very high temporal resolution—cannot be reached using Ca^{2+} imaging. Electrophysiological recordings and optical imaging of Ca^{2+} activity have, therefore, both advantages and limitations, and the electrode and genetic probes represent complementary tools to study the property of a neural circuit.

It should also be noted that Ca^{2+} imaging is not the only way of monitoring correlates of neural activity. Indicators reporting other aspects of neuronal function have been used in *Drosophila* as well. SynaptopHluorin is a genetically encoded sensor of neurotransmitter release and monitors the intravesicular change in pH value associated with vesicle fusion to the membrane (Miesenbock et al. 1998; Ng et al. 2002). Alternatively to fluorescence sensors, GFP-aequorin is a Ca^{2+}-sensitive bioluminescent photoprotein for probing the different dynamic aspects of Ca^{2+} signaling (Martin et al. 2007). In this case no excitation light is required and intracellular Ca^{2+} influx is directly reflected in photon emission. Although aequorin offers a large dynamic range, less than one photon is generated per aequorin molecule, so that the bioluminescence is very dim and difficult to visualize with good spatial resolution. A neuromodulator sensor called "Tango" has been recently developed, which can be used to visualize sites of neuromodulation (Inagaki et al. 2012).

7.4 Experimental Design

7.4.1 *Fly Strains*

A large collection of driver and responder strains are available from the stock centers for *D. melanogaster*.

Bloomington *Drosophila* Stock Center, Indiana University, USA.
(http://flystocks.bio.indiana.edu/)
Drosophila Genetic Resource Center, Kyoto Institute of Technology, Japan.
(http://kyotofly.kit.jp/cgi-bin/stocks/index.cgi)

7.4.2 *Preparation for Imaging the Brain*

The *Drosophila* brain is small enough to image large parts of its circuitry simultaneously. A major hindrance is, however, the pigmented cuticular exoskeleton that covers the fly's body surface. Parts of the cuticular head capsule have to be removed to provide optical access to the brain, which can be done by using fine forceps, splints, or a razor blade or an injection needle. An alternative approach is to completely expose the brain and use this isolated brain preparation (Shang et al. 2011).

Several types of imaging platforms such as a plastic cover slip or aluminum foil have been used to mount the fly (Fiala and Spall 2003; Joesch et al. 2008). An air-supported ball system has also been used for imaging the brain of a behaving fly under tethered condition (Chiappe et al. 2010; Kohatsu et al. 2011). In all cases, reducing movement artifacts by immobilizing the head is necessary to obtain high signal-to-noise ratio. Beeswax, silicone wax, and dental glue that are harmless to the animal are typically used to attach the fly head to the imaging platform.

7.4.3 Preparation for Imaging Sensory Organs

Because sensory neurons of insects are often associated with specialized cuticular structures such as bristles, hairs or sensillae, nondestructive imaging through the intact cuticle is in these cases required. Such noninvasive imaging is possible for sensory neurons located closely underneath the cuticle (Kamikouchi et al. 2009, 2010; Pelz et al. 2006). A fluorescent sensor protein with strong baseline fluorescence is useful to locate the sensory neuron of interest before the onset of the stimulus. Also in this case, immobilizing the body parts to be imaged is, of course, a critical step to get high signal-to-noise ratio.

7.4.4 Equipments and Data Analysis

In general the imaging equipment includes a fluorescence microscope, a light source for fluorophore excitation, and an emission light detector. Although the most powerful systems for Ca^{2+} imaging are probably laser-scanning microscopes, e.g., confocal laser-scanning or the two-photon microscopes, a simple combination of a standard upright epifluorescence microscope and a charge-coupled device (CCD) camera can for most applications deliver high-quality images. Once time series images of the GECI emission intensity are recorded, their fluorescence intensity can be analyzed off-line. Image processing, filtering, and averaging can all be used to eliminate noise and enhance calcium signals. For details of the imaging equipment and data analysis, see Part 3 ("Optical recording techniques").

7.4.5 Control of the Expression Level

Because GECIs are also Ca^{2+}-buffering proteins and therefore are potentially cytotoxic, optimizing the magnitude of expression to balance signal levels and cytotoxicity is required on a case-by-case basis. On the other hand, higher expression levels make the detection of fluorescence signals easier. Although a possible perturbation of endogenous Ca^{2+} signaling by the Ca^{2+}-buffering properties of GECIs has not been detected in *Drosophila* so far (Diegelmann et al. 2002; Jayaraman and Laurent 2007; Reiff et al. 2005), we recommend independent tests whether the functionality of the system you study is not affected by the expression of GECIs, e.g., through behavioral experiments. A couple of methods can be used to modify the level of expression. First, the use of homozygous versus heterozygous flies for both GAL4 and UAS transgenes alters the expression level. Two (or more) copies of a transgene typically result in higher levels of expression. Second, flies with different insertion sites, or number of UAS sites, can be used to vary the level of expression. More than one line for a single GECI is often available, such as UAS-Cameleon 2.1 (Fiala et al. 2002) and UAS-GCaMP3 (Tian et al. 2009), kept at the Bloomington *Drosophila* Stock Center. Third, the shift in temperature alters the activity of GAL4 in *Drosophila* (Duffy 2002). In flies maintained at 16 °C, GAL4 shows a minimal ability to activate transcription. As the temperature at which flies are raised is increased, the activity of GAL4 increases. 29 °C provides a balance between maximal GAL4 activity and minimal effects on fertility and viability of flies. This means that by altering the temperature, a wide range of expression levels the GECI can be achieved.

7.5 Examples of Applications

Here we exemplify how Ca^{2+} imaging in the brain or a sensory organ can be accomplished (Fig. 7.2). We monitored the neural responses in olfactory sensory neurons innervating the antennal lobe (AL), the primary olfactory center in the fly brain, and mechanosensory neurons in the Johnston's organ (JO), the auditory organ of the fly (Kamikouchi et al. 2010). The fly line UAS-Cameleon 2.1 (Diegelmann et al. 2002) was crossed to cell type-specific GAL4 lines, each of which expresses GAL4 in olfactory receptor neurons (Fig. 7.2a) or in the JO neurons (Fig. 7.2b), leading to restricted Cameleon 2.1 expression in the progeny. The flies were anesthetized on ice and then affixed to an imaging platform with a drop of glue. Images in the eYFP and eCFP channels were recorded simultaneously several seconds before, during, and after the stimuli as described previously (Fiala and Spall 2003). A time point is chosen before the onset of the stimulus to determine the baseline value for the intensities of the eYFP and the eCFP fluorescence (F_0) and the eYFP/eCFP ratio (R_0). The change in fluorescence relative to the baseline is calculated as $\Delta F/F_0$ for eYFP and eCFP, respectively. Likewise, the change in the ratio is calculated as $\Delta R/R_0$. This ratio change is indicative of alterations in the intracellular calcium concentration. It is clearly visible that in the case of olfactory receptor neurons, an odor stimulus causes an increase in intracellular Ca^{2+}, whereas in the case of JO neurons, sound is an adequate stimulus.

References

Benzer S (1967) Behavioral mutants of *Drosophila* isolated by countercurrent distribution. Proc Natl Acad Sci USA 58:1112–1119

Brand AH, Perrimon N (1993) Targeted gene expression as a means of altering cell fates and generating dominant phenotypes. Development 118:401–415

Chiang AS, Lin CY, Chuang CC et al (2011) Three-dimensional reconstruction of brain-wide wiring networks in *Drosophila* at single-cell resolution. Curr Biol 21:1–11

Chiappe ME, Seelig JD, Reiser MB et al (2010) Walking modulates speed sensitivity in *Drosophila* motion vision. Curr Biol 20:1470–1475

Dankert H, Wang L, Hoopfer ED et al (2009) Automated monitoring and analysis of social behavior in *Drosophila*. Nat Methods 6:297–303

Diegelmann S, Fiala A, Leibold C et al (2002) Transgenic flies expressing the fluorescence calcium sensor Cameleon 2.1 under UAS control. Genesis 34:95–98

Duffy JB (2002) GAL4 system in *Drosophila*: a fly geneticist's Swiss army knife. Genesis 34:1–15

Fiala A (2007) Olfaction and olfactory learning in *Drosophila*: recent progress. Curr Opin Neurobiol 17:720–726

Fiala A, Spall T (2003) In vivo calcium imaging of brain activity in *Drosophila* by transgenic cameleon expression. Sci STKE 2003:PL6

Fiala A, Spall T, Diegelmann S et al (2002) Genetically expressed cameleon in *Drosophila melanogaster* is used to visualize olfactory information in projection neurons. Curr Biol 12:1877–1884

Fischer JA, Giniger E, Maniatis T et al (1988) GAL4 activates transcription in *Drosophila*. Nature 332:853–856

Hasan MT, Friedrich RW, Euler T et al (2004) Functional fluorescent Ca2+ indicator proteins in transgenic mice under TET control. PLoS Biol 2:e163

Hay BA, Wolff T, Rubin GM (1994) Expression of baculovirus P35 prevents cell death in *Drosophila*. Development 120:2121–2129

Hayashi S, Ito K, Sado Y et al (2002) GETDB, a database compiling expression patterns and molecular locations of a collection of Gal4 enhancer traps. Genesis 34:58–61

Hendel T, Mank M, Schnell B et al (2008) Fluorescence changes of genetic calcium indicators and OGB-1 correlated with neural activity and calcium in vivo and in vitro. J Neurosci 28:7399–7411

Hendricks JC, Finn SM, Panckeri KA et al (2000) Rest in *Drosophila* is a sleep-like state. Neuron 25:129–138

Higashijima S, Masino MA, Mandel G et al (2003) Imaging neuronal activity during zebrafish behavior with a genetically encoded calcium indicator. J Neurophysiol 90:3986–3997

Hoyer SC, Eckart A, Herrel A et al (2008) Octopamine in male aggression of *Drosophila*. Curr Biol 18:159–167

Inagaki HK, de-Leon SB-T, Wong A et al (2012) Visualizing neuromodulation in vivo: TANGO-mapping of dopamine signaling reveals appetite control of sugar sensing. Cell 148:583–595

Jayaraman V, Laurent G (2007) Evaluating a genetically encoded optical sensor of neural activity using electrophysiology in intact adult fruit flies. Front Neural Circuits 1:3

Joesch M, Plett J, Borst A et al (2008) Response properties of motion-sensitive visual interneurons in the lobula plate of *Drosophila melanogaster*. Curr Biol 18:368–374

Kamikouchi A, Ito K (2007) The fruit fly - in vivo imaging by using GAL4-enhancer trap method. In: Miwa Y (ed) How to chose and use the fluorescent reagents for a successful experiment. YODOSHA, Tokyo

Kamikouchi A, Inagaki HK, Effertz T et al (2009) The neural basis of *Drosophila* gravity-sensing and hearing. Nature 458:165–171

Kamikouchi A, Wiek R, Effertz T et al (2010) Transcuticular optical imaging of stimulus-evoked neural activities in the *Drosophila* peripheral nervous system. Nat Protoc 5:1229–1235

Kerr R, Lev-Ram V, Baird G et al (2000) Optical imaging of calcium transients in neurons and pharyngeal muscle of *C. elegans*. Neuron 26:583–594

Kohatsu S, Koganezawa M, Yamamoto D (2011) Female contact activates male-specific interneurons that trigger stereotypic courtship behavior in *Drosophila*. Neuron 69:498–508

Lai SL, Lee T (2006) Genetic mosaic with dual binary transcriptional systems in *Drosophila*. Nat Neurosci 9:703–709

Lee T, Luo L (1999) Mosaic analysis with a repressible cell marker for studies of gene function in neuronal morphogenesis. Neuron 22:451–461

Luan H, Peabody NC, Vinson CR et al (2006) Refined spatial manipulation of neuronal function by combinatorial restriction of transgene expression. Neuron 52:425–436

Mank M, Santos AF, Direnberger S et al (2008) A genetically encoded calcium indicator for chronic in vivo two-photon imaging. Nat Methods 5:805–811

Martin JR, Rogers KL, Chagneau C et al (2007) In vivo bioluminescence imaging of Ca signalling in the brain of *Drosophila*. PLoS One 2:e275

Miesenbock G, De Angelis DA, Rothman JE (1998) Visualizing secretion and synaptic transmission with pH-sensitive green fluorescent proteins. Nature 394:192–195

Miyawaki A, Griesbeck O, Heim R et al (1999) Dynamic and quantitative Ca^{2+} measurements using improved cameleons. Proc Natl Acad Sci USA 96:2135–2140

Miyawaki A, Llopis J, Heim R et al (1997) Fluorescent indicators for Ca^{2+} based on green fluorescent proteins and calmodulin. Nature 388:882–887

Muto A, Ohkura M, Kotani T et al (2011) Genetic visualization with an improved GCaMP calcium indicator reveals spatiotemporal activation of the spinal motor neurons in zebrafish. Proc Natl Acad Sci USA 108:5425–5430

Nakai J, Ohkura M, Imoto K (2001) A high signal-to-noise Ca(2+) probe composed of a single green fluorescent protein. Nat Biotechnol 19:137–141

Nern A, Pfeiffer BD, Svoboda K et al (2011) Multiple new site-specific recombinases for use in manipulating animal genomes. Proc Natl Acad Sci USA 108:14198–14203

Ng M, Roorda RD, Lima SQ et al (2002) Transmission of olfactory information between three populations of neurons in the antennal lobe of the fly. Neuron 36:463–474

Ornitz DM, Moreadith RW, Leder P (1991) Binary system for regulating transgene expression in mice: targeting int-2 gene expression with yeast GAL4/UAS control elements. Proc Natl Acad Sci USA 88:698–702

Palmer AE, Giacomello M, Kortemme T et al (2006) Ca^{2+} indicators based on computationally redesigned calmodulin-peptide pairs. Chem Biol 13:521–530

Palmer AE, Tsien RY (2006) Measuring calcium signaling using genetically targetable fluorescent indicators. Nat Protoc 1:1057–1065

Pelz D, Roeske T, Syed Z et al (2006) The molecular receptive range of an olfactory receptor in vivo (*Drosophila melanogaster* Or22a). J Neurobiol 66:1544–1563

Pfeiffer BD, Jenett A, Hammonds AS et al (2008) Tools for neuroanatomy and neurogenetics in *Drosophila*. Proc Natl Acad Sci USA 105:9715–9720

Pfeiffer BD, Ngo TT, Hibbard KL et al (2010) Refinement of tools for targeted gene expression in *Drosophila*. Genetics 186:735–755

Potter CJ, Tasic B, Russler EV et al (2010) The Q system: a repressible binary system for transgene expression, lineage tracing, and mosaic analysis. Cell 141:536–548

Reiff DF, Ihring A, Guerrero G et al (2005) In vivo performance of genetically encoded indicators of neural activity in flies. J Neurosci 25:4766–4778

Scheer N, Campos-Ortega JA (1999) Use of the Gal4-UAS technique for targeted gene expression in the zebrafish. Mech Dev 80:153–158

Seelig JD, Chiappe ME, Lott GK et al (2010) Two-photon calcium imaging from head-fixed *Drosophila* during optomotor walking behavior. Nat Methods 7:535–540

Shang Y, Haynes P, Pirez N et al (2011) Imaging analysis of clock neurons reveals light buffers the wake-promoting effect of dopamine. Nat Neurosci 14:889–895

Shimada T, Kato K, Kamikouchi A et al (2005) Analysis of the distribution of the brain cells of fruit fly by an automatic cell counting algorithm. Physica A Stat Phys 350:144–149

Spradling AC, Rubin GM (1982) Transposition of cloned P elements into *Drosophila* germ line chromosomes. Science 218:341–347

Tian L, Hires SA, Mao T et al (2009) Imaging neural activity in worms, flies and mice with improved GCaMP calcium indicators. Nat Methods 6:875–881

Venken KJ, Simpson JH, Bellen HJ (2011) Genetic manipulation of genes and cells in the nervous system of the fruit fly. Neuron 72:202–230

von Philipsborn AC, Liu T, Yu JY et al (2011) Neuronal control of *Drosophila* courtship song. Neuron 69:509–522

Vosshall LB, Stocker RF (2007) Molecular architecture of smell and taste in *Drosophila*. Annu Rev Neurosci 30:505–533

Wang JW, Wong AM, Flores J et al (2003) Two-photon calcium imaging reveals an odor-evoked map of activity in the fly brain. Cell 112:271–282

Wu JS, Luo L (2006) A protocol for mosaic analysis with a repressible cell marker (MARCM) in *Drosophila*. Nat Protoc 1:2583–2589

Yao KM, White K (1994) Neural specificity of elav expression: defining a *Drosophila* promoter for directing expression to the nervous system. J Neurochem 63:41–51

Chapter 8
Controlling Behavior Using Light to Excite and Silence Neuronal Activity

Ali Cetin and Shoji Komai

Abstract The number of reports involving the new tools of optogenetics is increasing exponentially to yield detailed insights into anatomical, physiological, and pathological issues. These tools help us to tackle major questions regarding the function of neural circuits in the mammalian brain, which possesses uncountable combinations of neurons. Moreover, rapid progress in diverse collaborations between optogenetics and optical imaging technologies will allow us to analyze, simultaneously, the activities of multiple neurons and glial cells. As well as activity analysis, optogenetics is developing rapidly to support the analysis of stimulation in neuronal function. We can now stimulate multiple cell types independently using selective molecular tools, such as promoters and gene delivery systems. In addition, optical properties also help us to discriminate among subpopulations of cells in neuronal networks. The use of light to study the brain has proved to be a remarkably fruitful strategy, and indeed optogenetics has given us a green light for the future.

Keywords Behavioral control • Gene delivery strategies • Neuronal activity • Optogenetics • Rhodopsin

8.1 Introduction

Optogenetics has been applied to various neurobiological questions to elucidate in detail the mechanisms underlying behavior. Optogenetics was established in 2006 by Karl Deisseroth and colleagues, who expressed a photoactivatable protein called

A. Cetin
Allen Institute for Brain Science, Seattle WA, USA

S. Komai (✉)
Nara Institute of Science and Technology, Ikoma, Nara, Japan
e-mail: skomai@bs.naist.jp

channelrhodopsin-2 (ChR2) in particular neurons to stimulate them specifically by light. ChR2 thus allows us to control neurons with high spatiotemporal resolution. As with calcium imaging, optogenetics provides a molecular tool with which single neurons, or populations of neurons, within a network of interest can be studied.

8.1.1 Optogenetics

Trillions of synapses connecting neurons are included in the whole network of the mammalian brain (Luo et al. 2008). The components of this huge network, such as neurons and glias, are well characterized in all areas of the brain. These components build up not only a relatively simple system but also an incredibly complex system, corresponding to sensory or motor processing and higher cognitive functions, such as the circuits that mediate vision, motor movements, breathing/respiration, and sleep/wake architecture. At both levels of complexity, the relative contributions of individual cells and the synaptic connections between them should be elucidated to understand the mechanisms of information processing in the brain. However, precise manipulation of the activities of the network has been extremely challenging for traditional electrophysiological, pharmacological, and genetic methods (Luo et al. 2008; Carter and Shieh 2010). While numerous successes have been reported with these classical methods, we have faced considerable obstacles to achieving spatiotemporal precision in the study of neural circuits in vivo. Conventional electrical and physical techniques are spatially imprecise, since surrounding cells are also stimulated or inhibited. Pharmacological and genetic methods yield better data than electrical and physical approaches in terms of spatial selectivity, but temporal resolution is still insufficient for single action potentials. To overcome these limitations, optogenetics has been developed as a new set of tools to precisely stimulate (Boyden et al. 2005; Zhang et al. 2008; Berndt et al. 2009; Lin et al. 2009; Gunaydin et al. 2010; Li et al. 2005; Nagel et al. 2003), inhibit neural activity (Chow et al. 2010; Gradinaru et al. 2010; Zhao et al. 2008; Gradinaru et al. 2008; Zhang et al. 2007a, b), or alter biochemical activity in specific cells (Airan et al. 2009; Oh et al. 2010) with high temporal precision and rapid reversibility. These tools are activated by light ("opto-") and are genetically encoded ("-genetics") to give us specific control over particular populations of cells in vitro and in vivo (Fig. 8.1) (Zhang et al. 2006; Gradinaru et al. 2007; Zhang et al. 2007a, b; Zhang et al. 2010; Cardin et al. 2010). The precise manipulation that these new tools permit has facilitated further progress in elucidating structural and functional aspects of neural circuits.

8 Controlling Behavior Using Light to Excite and Silence Neuronal Activity

Fig. 8.1 Concept of optogenetics. Rhodopsin molecules (here, ChR2, shown on the membrane in the foreground and on the neuron just behind it) are expressed in a particular type of neurons using molecular tools such as viral vectors or transgenic animals. When a neuron is stimulated by light, cations (*green*) enter the cell, which results in excitement of the ChR2-expressing neurons (shown as bright spots in the dendrites). Optical fibers, through which a stimulus light source is introduced, are available for in vivo experiments

8.2 Revealing Dynamism Based on Physiological and Pathological Neuronal Networks

8.2.1 Retina

Since retinal degeneration is characterized by the progressive loss of rod and cone photoreceptors, targeting ChR2 to retinal ganglion cells (RGCs) or bipolar cells could theoretically restore light sensitivity to mammalian models of retinal degeneration in which most rods and cones are genetically ablated (Tomita et al. 2009). Expression of ChR2 in either RGCs (Bi et al. 2006) or bipolar cells (Lagali et al. 2008) could indeed restore visual sensitivity in mice with mutations causing rod and cone degeneration, as well as in rat models of retinal degeneration caused either by genetic mutation (Tomita et al. 2010) or toxic light exposure (Tomita et al. 2009).

Halorhodopsin (NpHR) could potentially be used in these cells to restore vision, because photons naturally cause hyperpolarization of photoreceptors. Interestingly, expression of NpHR in the inner retinal layer could restore OFF responses, whereas combined expression of NpHR and ChR2 in RGCs could cause them to respond as ON, OFF, or even ON-OFF cells depending on the wavelength of light used (Zhang et al. 2009). The expression of "enhanced" NpHR (eNpHR) in light-insensitive cone photoreceptors could substitute for the native phototransduction cascade and restore light sensitivity in two mouse models of retinal degeneration (Busskamp et al. 2010). Importantly, this treatment leads to normal activity in cone photoreceptors and RGCs in response to yellow-light stimulation and allows mice to respond to changes in light intensity and the direction of motion of visual stimulation.

Even in human retinas, eNpHR expression was nontoxic and could rescue light-insensitive human photoreceptors ex vivo (Busskamp et al. 2010). These results demonstrate that optogenetics may be useful for treating various forms of blindness in humans. Optogenetics may also be utilized to establish various prosthetic devices, such as specialized glasses that increase light intensity, which have been proposed to enhance environmental visual stimuli specifically for ChR2- or eNpHR-transduced neurons in the retina (Cepko 2010).

8.2.2 Reflex and Innate Behavior

Breathing/Respiration

Brainstem or spinal cord injury may result in paralysis, which leads to the inability to breathe in severe cases. Although the neuronal mechanism of respiration is not sufficiently well understood to restore function to damaged circuits, ChR2-mediated photostimulation of motor neurons has recently been reported to recover respiratory diaphragmatic motor activity (Alilain et al. 2008). In the brainstem, stimulation of the retrotrapezoid nucleus produced long-lasting activation of breathing (Abbott et al. 2009a), whereas stimulation of the ventrolateral medulla increased sympathetic nerve activity and blood pressure (Abbott et al. 2009b). These studies indicate that neural and non-neural cell types in the brainstem and spinal cord contribute to the regulation of a central autonomic process.

Sleep/Wake Circuitry

The sleep/wake cycle is one of the most well-defined behaviors whose underlying principles have been dissected by optogenetic tools. The first use of optogenetics to study this system used lentivirus-mediated gene delivery to target ChR2 to hypocretin-expressing neurons in the lateral hypothalamus (Adamantidis et al. 2007). Dysregulation of the hypocretin system, either of the peptide or its receptor, causes the sleep disorder narcolepsy. Photostimulation on ChR2-expressing hypocretin neurons increased the probability of sleep/wake transitions (Adamantidis et al. 2007) and increased neuronal activity in downstream wake-promoting nuclei (Carter et al. 2009). It has also been shown that optogenetic stimulation of the locus coeruleus (LC) produces immediate sleep-to-wake transitions, whereas optogenetic inhibition causes a decrease in wakefulness (Carter et al. 2010). In addition, tonic stimulation of the LC led the mice to a cataplexy-like state, which is a sign of narcolepsy.

8.2.3 Motor Behavior

The selective loss of dopaminergic neurons in the substantia nigra pars compacta leads to a severe neurodegenerative disorder known as Parkinson's disease, characterized by muscle rigidity and uncoordinated physical movements. Two major output routes from striatal neurons, known as the direct and indirect pathways, are thought to be related to this disease. The indirect pathway output neurons express a specific dopamine receptor called dopamine receptor type 2 (D2R), a G-protein-coupled receptor that inhibits adenylate cyclase and thus suppresses calcium signaling. Stimulation of D2R by dopamine leads to suppression of the indirect pathway. Under normal circumstances, the indirect pathway suppresses the inhibitory output of the external capsule of globus pallidus (GPe), which is involved in suppressing the subthalamic nucleus (STN). It is generally thought that losing dopamine results in increased indirect pathway activity, due to the loss of D2R-mediated suppression of this pathway. When GPe can no longer function as a major inhibitor of the STN, the ensuing net increase in STN activity may explain the motor symptoms of Parkinson's disease. Optogenetics has recently clarified that activation of the indirect pathway indeed mimics the parkinsonian state (Kravitz et al. 2010). In this study, the relative contributions of dopamine D1-receptor- and D2-receptor-expressing neurons in the striatum were investigated by selectively targeting each type with ChR2. Stimulation of D1-expressing neurons in the striatum reduced parkinsonian symptoms in a mouse model of the disease, whereas stimulation of D2-expressing neurons—which mimics dopamine depletion of indirect pathway—caused symptoms in wild-type mice.

In addition to deepening our understanding of Parkinson's disease circuitry, another recent optogenetics study shed light on how one of the traditional treatments for this disease may be relieving the symptoms. A crude technique called deep-brain electrical stimulation within STN causes a reversal of Parkinson's disease symptoms. To understand the underlying principles of this phenomenon, optogenetic probes were used to systematically stimulate or inhibit a mixture of distinct circuit elements containing neurons, glia, and fiber projections in the STN of freely moving rodent models of Parkinson's disease (Gradinaru et al. 2009). When the excitatory afferent axons projecting to the STN—such as those of motor neurons in

M1—were optogenetically stimulated, the parkinsonian symptoms were relieved. This study demonstrated that the excitatory afferent regulation of STN is at the heart of the beneficial outcome of deep-brain stimulation. Taken together, these results promote our understanding of the functional connections within the basal ganglia and may contribute to current therapeutic strategies to ameliorate parkinsonian motor deficits (Bernstein et al. 2008).

8.2.4 Memory Formation and Reinforcement

The neural circuits based on reinforcement have been well characterized with optogenetic tools. Stimulation of dopaminergic neurons in the ventral tegmental area co-released glutamate as well as dopamine into the nucleus accumbens, demonstrating that mesolimbic reward signaling involves glutamatergic transmission (Tecuapetla et al. 2010; Stuber et al. 2010). Optical stimulation of α1-adrenergic receptors in the nucleus accumbens, but not of β2-adrenergic receptors, led to a robust increase in place preference during conditioning (Airan et al. 2009). In addition, the relative contributions of distinct tonic versus phasic activity patterns in participating brain structures, as well as the relative contributions of modulatory systems with various neurotransmitters, are unknown (Stuber 2010). By means of optogenetics, selective phasic photostimulation of dopaminergic neurons in the ventral tegmental area was shown to be sufficient to establish association learning, whereas tonic activation was not (Tsai et al. 2009).

8.2.5 Anxiety and Aggression

Even a fairly deep part of the brain, the ventromedial hypothalamus (VMH), which is thought to be closely related to instinctive behaviors, could be photostimulated through an implanted optical fiber. Optogenetic, but not electrical, stimulation of neurons in the VMH ventrolateral subdivision (VMHvl) causes male mice to attack both females and inanimate objects, as well as males (Lin et al. 2011).

Another group, using of ChR2-assisted circuit mapping in amygdala slices and cell-specific viral tracing, has reported that protein kinase C-d (PKC-d)1 neurons inhibit output neurons in the medial central amygdala (CEm) and also make reciprocal inhibitory synapses with PKC-d2 neurons in the lateral subdivision of the central amygdala (CEl). These results, together with behavioral data, define an inhibitory microcircuit in CEl that gates CEm output to control the level of conditioned freezing (Haubensak et al. 2010). In another study, specific optogenetic stimulation of oxytocinergic axons in the amygdala was shown to reduce freezing responses in fear-conditioned rats, illuminating the mechanisms by which oxytocin modifies emotional circuitry in a positive manner (Knobloch et al. 2012).

8.2.6 Balance Between Excitatory and Inhibitory Networks

Various oscillatory fluctuations have been observed in the cortex, which is thought to be associated with particular behavior. Cortical gamma oscillations (30–100 Hz) have been well elucidated, but the neural basis of these rhythms, and their role in animal behavior, remain unknown. ChR2-mediated photostimulation of parvalbumin (PV)-expressing interneurons amplified gamma oscillations, whereas eNpHR-mediated photoinhibition suppressed them (Sohal et al. 2009; Cardin et al. 2009). Furthermore, γ-frequency modulation of excitatory input enhanced signal transduction in cortical regions, reducing circuit noise and amplifying circuit signals. These studies provide the first causal evidence that distinct network activity states can be induced in vivo by cell-type-specific activation of PV neurons, and also suggest a potential mechanism for the altered γ-frequency synchronization and cognition in schizophrenia and autism (Sohal et al. 2009; Cardin et al. 2009). In another recent study, optogenetic stimulation of layer VI excitatory neurons was shown to reduce firing within the upper layers of the mouse visual cortex. This suggests that activation of layer VI excitatory neurons plays an essential role in gain control within cortical sensory networks (Olsen et al. 2012). In addition to the neurological studies described above, optogenetic probes have been used to investigate many other aspects of health and disease, including associative fear memory (Haubensak et al. 2010; Ciocchi et al. 2010), epilepsy (Tonnesen et al. 2009), and the blood oxygen level-dependent (BOLD) effect during functional magnetic resonance imaging (Lee et al. 2010).

8.3 Technical Aspects

8.3.1 Molecular Aspects

Although there are notable exceptions, the most commonly used optogenetic probes are gene-engineered versions of natural opsins, which are light-sensitive membrane proteins through which ions are translocated in response to light stimulation at specific wavelengths (Kramer et al. 2009). These probes can be utilized either to excite the cells, to inhibit their activity, or to change intracellular signaling (Fig. 8.2).

Probes for Stimulating Neurons

ChR2 is a nonspecific cation channel naturally expressed in the alga *Chlamydomonas reinhardtii* (Nagel et al. 2003). On absorbing blue light at an absorption peak of 480 nm, ChR2 undergoes a conformational change from the all-*trans*-retinal chromophore complex to 13-*cis*-retinal (Bamann et al. 2008). This switch causes a

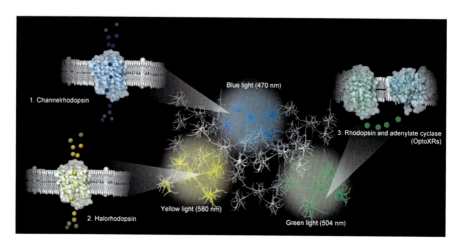

Fig. 8.2 Examples of optogenetics. Membrane excitability is manipulated using various optogenetic tools. ChR2 and NpHR allow us to excite and inhibit particular neurons, respectively, and thereby to control excitability by applying different excitation wavelengths even when these modified neurons coexist in a particular region of interest. Similarly, intracellular signaling can also be modified with light-triggered G-proteins, "OptoXRs." The colored balls in *blue*, *yellow*, and *green* indicate cation, anion, and intracellular signal, respectively

subsequent conformational change in the channel protein to open the pore, allowing various cations—such as H^+, Na^+, K^+, and Ca^{2+}—to flow (Nagel et al. 2003; Bamann et al. 2008). ChR2 has several features that make it particularly attractive as a neuroscience probe to depolarize neurons. First, the channel can be activated very rapidly and closes quickly upon light offset. 13-*cis*-retinal relaxes back to the all-*trans* form within milliseconds, closing the pore and stopping the flow of ions into or out of the cell. Therefore, single action potentials can be generated with a brief pulse of blue light, without any accompanying inappropriate effects of stimulation. Second, retinal is already present in most vertebrate cells in the form of vitamin A, which allows the ChR2 apoprotein to become a light-sensitive holoprotein. Therefore, the extraretinal is not needed when ChR2 is used in vertebrate neural systems, although exogenous application is necessary for invertebrate systems. Finally, as a genetically encoded protein, ChR2 permits cell-specific targeting with defined promoter and enhancer elements. Taken together, these properties of ChR2 allow researchers to stimulate particular neurons of interest with millisecond-level temporal resolution (Boyden et al. 2005; Li et al. 2005).

Recently, various opsins have been developed that can help us to analyze brain functions effectively. The red-shifted opsin VChR1, for example, is useful when an additional wavelength of light is required for excitation, while step-function opsins (SFOs) can inject a stable step current to allow a positive shift in membrane potential for up to 30–60 s with a single brief pulse, after which channel closure can be triggered by a brief exposure to yellow light.

Probes for Inhibiting Neurons

NpHR was the first molecule shown to inhibit neural activity (Zhang et al. 2007a, b; Han and Boyden 2007) and is naturally expressed in the halobacterium *Natronomonas pharaonis* (Schobert and Lanyi 1982; Bamberg et al. 1993). NpHR is a light-driven pump, which actively pumps Cl ions into cells in response to yellow light at the peak absorption wavelength of 570 nm. Like ChR2, NpHR utilizes retinal as its chromophore and therefore can also be used in vertebrate systems without extra cofactors. Substantial mutagenesis was required to achieve high levels of expression in neurons. Enhanced halorhodopsin (eNpHR), a second-generation NpHR, possesses an endoplasmic reticulum export signal and thus displays improved translocation to the plasma membrane (Zhao et al. 2008; Gradinaru et al. 2008). Third-generation constructs (eNpHR3.0) have been generated to improve photocurrent increasing in membrane hyperpolarization over eNpHR, with additional membrane-trafficking sequences (Gradinaru et al. 2010). Other proteins related to bacteriorhodopsins have been discovered recently and can be used to inhibit neural activity in response to light. Unlike NpHR, these proteins function as light-driven proton pumps. Archaerhodopsin-3 (Arch) proteins are derived from *Halorubrum sodomense* and allow near-100 % silencing of neurons in vivo in response to yellow light, with an efficiency comparable to that of eNpHR3.0 (Chow et al. 2010). Two other bacterial rhodopsins, Mac proteins from *Leptosphaeria maculans* and bacteriorhodopsin (BR) from *Halobacterium salinarum* (and its enhanced, second-generation derivative, eBR), allow silencing of neurons in response to blue-green light (Chow et al. 2010; Gradinaru et al. 2010). Therefore, hyperpolarizing optogenetic tools now exist that respond to blue– green and yellow light, allowing for combinatorial dissection of two neural subtypes in the same preparation. High-throughput genomic screens should reveal additional channels and thereby increase the diversity of inhibitory optogenetic tools for future use.

Interestingly, in a recent set of experiments, a proton-pumping archeorhodopsin was shown to allow high-speed imaging of individual action potentials. When a single amino acid residue was mutated, the resulting structure prevented proton flow across the membrane, allowing the channel to be used solely as an indicator. This genetically encoded voltage indicator exhibited an approximately tenfold improvement in sensitivity and speed over existing protein-based voltage indicators, with a roughly linear twofold increase in brightness between −150 and +150 mV and a sub-millisecond response time (Kralj et al. 2011). Therefore, optogenetics allows not only selective control of neural circuitry but also readout of its signals.

Probes for Manipulation of Intracellular Signaling

Neurons are also modulated by intracellular signaling events, initiated by cell surface receptors that culminate in a change in neuronal electrical activity instead of electrical signals through ion channels, as well as changes in secondary messenger pathways leading to gene expression and downstream protein cascades. Because rhodopsins are members of the GPCR family, it is theoretically possible to design synthetic rhodopsin-GPCR chimeras that combine the light-responsive elements of rhodopsin with the machinery of biochemical signaling of specific GPCRs (Kramer et al. 2009; Karnik et al. 2003; Kim et al. 2005).

Optogenetic Gene Delivery Strategies

The commonest method of delivering optogenetic transgenes into the nervous system is to infect cells with a self-inactivating virus, typically a lentivirus or adeno-associated virus (AAV), that contains the transgene of interest driven by a short promoter or enhancer element (Luo et al. 2008). Cell-type-specific promoters that have been used to drive optogenetic transgenes include EF1a (strong, ubiquitous expression) (Tsai et al. 2009; Cardin et al. 2009), CAMKII-α (expression limited to excitatory neurons) (Boyden et al. 2005; Gradinaru et al. 2009), synapsin I (limited to neurons) (Zhang et al. 2007a, b), and GFAP (limited to astrocytes) (Gradinaru et al. 2009). Several other promoters are useful for targeting to specific cell types in the brain, such as the ppHcrt promoter that targets hypocretin-expressing neurons in the lateral hypothalamus (Adamantidis et al. 2007), the oxytocin promoter that targets oxytocin peptide hormone-releasing neurons of the hypothalamus (Knobloch et al. 2012), and the synthetic PRSx8 promoter, which targets noradrenergic and adrenergic neurons that express dopamine beta hydroxylase (Abbott et al. 2009a; Abbott et al. 2009b). In utero electroporation is also useful for introducing optogenetic transgenes at specific developmental stages. Transgenes can be delivered to specific cortical layers of the brain by electroporating mice at embryonic day E12.5 (layers V and VI), E13.5 (layer IV), or E15.5 (layers II and III) (Zhang et al. 2010). Several studies have used this approach to deliver ChR2 to specific cortical layers for subsequent photostimulation when the mice reach adulthood (Gradinaru et al. 2007; Hull et al. 2009; Adesnik and Scanziani 2010).

Viral, transgenic, and in utero electroporation strategies can be combined to overcome the weak transcriptional activity of most endogenous promoters (Zhang et al. 2010). AAV vectors expressing Cre-dependent transgene cassettes under the control of strong, ubiquitous promoters such as EF1a have been developed which allow us to utilize numerous transgenic mouse lines from individual labs, such as GENSAT (Gong et al. 2007) and the Allen Institute for Brain Science (Madisen et al. 2010), that express Cre recombinase in specific cell types. Indeed, many optogenetic studies have capitalized on these excellent systems (Tsai et al. 2009; Cardin et al. 2009). Additionally, it is possible to specify the expression of optogenetic transgenes using anatomical-based cell targeting (Gradinaru et al. 2010). Rabies viral vectors or proteins such as wheat germ agglutinin or tetanus toxin fragment C help the gene of interest or Cre recombinase to be transported anterogradely and retrogradely (Luo et al. 2008; Gradinaru et al. 2010; Zhang et al. 2010; Osakada et al. 2011; Wall et al. 2010). Thus, it is possible to restrict the expression of optogenetic transgenes in a particular neural circuit even if its cells do not express unique genetic regulatory elements. Finally, cellular indicators of functional activity, including the immediate early genes zif268 (egr1), c-fos, and arc, help us to regulate the activated network relating to particular behaviors (Covington et al. 2010; Liu et al. 2012). The combination of these technologies provides scientists with multiple strategies for expressing optogenetic probes in specific neural networks, and the technologies also offer greater flexibility to express the probes in various model animals, such as rats and primates.

8.3.2 Optical Aspects

To manipulate neural activities in a specific population of cells, it is necessary to deliver light properly. A light delivery system for in vitro cultured neurons or brain slices can be established using conventional light sources such as halogen/xenon arc lamps, light-emitting diodes (LEDs), and lasers, all of which can be directly built into the light path of a microscope (Zhang et al. 2010). In vivo light delivery, however, is more challenging because surgical implants of optical fibers are required to place them stereotaxically around targeted regions. Light needs to be delivered very close to target regions because brain tissue scatters light exponentially, with only 10 % of light intensity remaining at a distance of 500 µm from the light source (Adamantidis et al. 2007; Aravanis et al. 2007). Regarding freely moving animals, the delivery device must be light enough to be carried easily and should not interfere with natural behaviors. At present, the commonest method for delivering light in vivo is to implant a guide cannula to place a fiber optical cable (Fig. 8.1) (Aravanis et al. 2007). This procedure allows cells in deep-brain structures to be targeted and is applicable for mice with up to 300-µm diameter fibers and for rats with up to 400 µm fibers. Optical fibers are typically connected to a laser diode, although it is also possible to connect them to an LED. To modulate the neural activities of superficial cortical neurons, small LEDs can be mounted above a glass over a cranial window (Gradinaru et al. 2007; Huber et al. 2008, Wentz et al. 2011). Recently, an array of optical fibers or an electrocorticogram (ECoG) equipped with LEDs has been established to modulate simultaneously multiple sites in the brain (Bernstein and Boyden 2011; Sawadsaringkarn et al. 2012). These technologies help us to analyze information processing throughout the brain.

The two-photon microscope is one of the most widespread tools to observe neurons of interest, because it allows deeper areas of the brain to be visualized than does conventional microscopy. However, its adoption for optogenetics presents technical difficulties. The typical diffraction-limited focal volumes required to achieve the conditions of multiphoton excitation are about 1/1,000 of the volume of a typical cell body. This strategy enables us to activate only a tiny fraction of the available channels on the cell's plasma membrane, but would yield a degree of stimulation insufficient to depolarize most cells to the action potential threshold. Trying to increase this focal volume by reducing the effective numerical aperture of the optical system fails to overcome the problem because the lateral and axial dimensions of the excited focal spot are coupled: a tenfold decrease in lateral resolution corresponds to a 100-fold loss of axial resolution. A simple solution to this resolution versus effectiveness tradeoff is to move the stimulation spot very rapidly across the cell membrane, integrating the cumulative effect of many locations (Rickgauer and Tank 2009) for two-photon ChR2 neural stimulation with a 30 ms spiral scan.

To identify functional connections in the real brain, the cell-type- and site-specific causal controls provided by optogenetics and fMRI in mice have been combined to test the linearity of BOLD signals driven by locally induced excitatory activity (Lee et al. 2010; Kahn et al. 2011; Desai et al. 2011). This strategy helps us to estimate how linear the response to sensory stimuli is, which is essential for the design and interpretation of in vivo fMRI experiments.

Additional information about optogenetics, including technical details about genes and light delivery systems, can be found in other excellent reviews and protocols (Zhang et al. 2006; Cardin et al. 2010). More information about available optogenetic transgenes can be found at the Optogenetics Resource Center Web page (http://www.stanford.edu/group/dlab/optogenetics/) maintained by the laboratory of Karl Deisseroth. Details of tools for gene and light delivery are available on the Web page (http://syntheticneurobiology.org/protocols) maintained by the laboratory of Ed Boyden.

8.4 Conclusion

Although optogenetics has been established for only 6 years, the number of reports exploiting these tools is increasing exponentially to provide further answers to anatomical, physiological, and pathological issues. These tools allow us to address the complexity of neural circuits, for not only in vitro but also in vivo studies. However, there remains much scope for further refinement that would allow us, for example, to stimulate multiple cell types at the same time by introducing various mutations in a particular domain, as is the case with the multiple available versions of green fluorescent protein. It will also be desirable to increase the conductance of various channels so that less light stimulation is necessary, to avoid the expected side effects of heat on the cell. Finally, it should also be possible to record patterns or sequences of neural firing and then use these recordings to mimic the recorded neural activities with light pulses. Regarding the therapeutic use of optogenetics, it will be necessary to develop safe and reversible gene delivery strategies and light delivery devices that are adaptable to human patients. The use of light to study the brain has proved to be a remarkably fruitful strategy, and indeed optogenetics has given us a green light for the future.

Acknowledgements We thank all members of our laboratories for useful discussion and support. Fig. 8.1 was adopted from the front page of BioGARAGE, published on March 2012 by Leave a Nest Co., Ltd. The images in both figures were designed by Science Graphics Co., Ltd.

References

Abbott SB, Stornetta RL, Fortuna MG et al (2009a) Photostimulation of retrotrapezoid nucleus phox2b-expressing neurons in vivo produces long-lasting activation of breathing in rats. J Neurosci 29:5806–5819

Abbott SB, Stornetta RL, Socolovsky CS et al (2009b) Photostimulation of channelrhodopsin-2 expressing ventrolateral medullary neurons increases sympathetic nerve activity and blood pressure in rats. J Physiol 587:5613–5631

Adamantidis AR, Zhang F, Aravanis AM et al (2007) Neural substrates of awakening probed with optogenetic control of hypocretin neurons. Nature 450:420–424

Adesnik H, Scanziani M (2010) Lateral competition for cortical space by layer-specific horizontal circuits. Nature 464:1155–1160

Airan RD, Thompson KR, Fenno LE et al (2009) Temporally precise in vivo control of intracellular signalling. Nature 458:1025–1029

Alilain WJ, Li X, Horn KP et al (2008) Light-induced rescue of breathing after spinal cord injury. J Neurosci 28:11862–11870

Aravanis AM, Wang LP, Zhang F et al (2007) An optical neural interface: in vivo control of rodent motor cortex with integrated fiberoptic and optogenetic technology. J Neural Eng 4:S143–56

Bamann C, Kirsch T, Nagel G et al (2008) Spectral characteristics of the photocycle of channelrhodopsin-2 and its implication for channel function. J Mol Biol 375:686–694

Bamberg E, Tittor J, Oesterhelt D (1993) Light-driven proton or chloride pumping by halorhodopsin. Proc Natl Acad Sci USA 90:639–643

Berndt A, Yizhar O, Gunaydin LA et al (2009) Bi-stable neural state switches. Nat Neurosci 12:229–234

Bernstein JG, Han X, Henninger MA et al (2008) Prosthetic systems for therapeutic optical activation and silencing of genetically-targeted neurons. Proc Soc Photo Opt Instrum Eng 6854:68540H

Bernstein JG, Boyden ES (2011) Optogenetic tools for analyzing the neural circuits of behavior. Trends Cogn Sci 15:592–600

Bi A, Cui J, Ma YP et al (2006) Ectopic expression of a microbial-type rhodopsin restores visual responses in mice with photoreceptor degeneration. Neuron 50:23–33

Boyden ES, Zhang F, Bamberg E et al (2005) Millisecond-timescale, genetically targeted optical control of neural activity. Nat Neurosci 8:1263–1268

Busskamp V, Duebel J, Balya D et al (2010) Genetic reactivation of cone photoreceptors restores visual responses in retinitis pigmentosa. Science 329:413–417

Cardin JA, Carlen M, Meletis K et al (2010) Targeted optogenetic stimulation and recording of neurons in vivo using cell-type-specific expression of Channelrhodopsin-2. Nat Protoc 5:247–254

Cardin JA, Carlen M, Meletis K et al (2009) Driving fast-spiking cells induces gamma rhythm and controls sensory responses. Nature 459:663–667

Carter ME, Adamantidis A, Ohtsu H et al (2009) Sleep homeostasis modulates hypocretin-mediated sleep-to-wake transitions. J Neurosci 29:10939–10949

Carter ME, Yizhar O, Chikahisa S et al (2010) Tuning arousal with optogenetic modulation of locus coeruleus neurons. Nat Neurosci 13:1526–1533

Carter M, Shieh JC (2010) Guide to research techniques in neuroscience. Elsevier/Academic, Amsterdam/Boston

Cepko C (2010) Neuroscience. Seeing the light of day. Science 329:403–404

Chow BY, Han X, Dobry AS et al (2010) High-performance genetically targetable optical neural silencing by light-driven proton pumps. Nature 463:98–102

Ciocchi S, Herry C, Grenier F et al (2010) Encoding of conditioned fear in central amygdala inhibitory circuits. Nature 468:277–282

Covington HE 3rd, Lobo MK, Maze I et al (2010) Antidepressant effect of optogenetic stimulation of the medial prefrontal cortex. J Neurosci 30:16082–16090

Desai M, Kahn I, Knoblich U et al (2011) Mapping brain networks in awake mice using combined optical neural control and fMRI. J Neurophysiol 105:1393–1405

Gong S, Doughty M, Harbaugh CR et al (2007) Targeting Cre recombinase to specific neuron populations with bacterial artificial chromosome constructs. J Neurosci 27:9817–9823

Gradinaru V, Mogri M, Thompson KR et al (2009) Optical deconstruction of parkinsonian neural circuitry. Science 324:354–359

Gradinaru V, Thompson KR, Deisseroth K (2008) eNpHR: a Natronomonas halorhodopsin enhanced for optogenetic applications. Brain Cell Biol 36:129–139

Gradinaru V, Thompson KR, Zhang F et al (2007) Targeting and readout strategies for fast optical neural control in vitro and in vivo. J Neurosci 27:14231–14238

Gradinaru V, Zhang F, Ramakrishnan C et al (2010) Molecular and cellular approaches for diversifying and extending optogenetics. Cell 141:154–165

Gunaydin LA, Yizhar O, Berndt A et al (2010) Ultrafast optogenetic control. Nat Neurosci 13:387–392

Han X, Boyden ES (2007) Multiple-color optical activation, silencing, and desynchronization of neural activity, with single-spike temporal resolution. PLoS One 2:e299

Haubensak W, Kunwar PS, Cai H et al (2010) Genetic dissection of an amygdala microcircuit that gates conditioned fear. Nature 468:270–276

Huber D, Petreanu L, Ghitani N et al (2008) Sparse optical microstimulation in barrel cortex drives learned behaviour in freely moving mice. Nature 451:61–64

Hull C, Adesnik H, Scanziani M (2009) Neocortical disynaptic inhibition requires somatodendritic integration in interneurons. J Neurosci 29:8991–8995

Kahn I, Desai M, Knoblich U et al (2011) Characterization of the functional MRI response temporal linearity via optical control of neocortical pyramidal neurons. J Neurosci 31:15086–15091

Karnik SS, Gogonea C, Patil S et al (2003) Activation of G-protein-coupled receptors: a common molecular mechanism. Trends Endocrinol Metab 14:431–437

Kim JM, Hwa J, Garriga P et al (2005) Light-driven activation of beta 2-adrenergic receptor signaling by a chimeric rhodopsin containing the beta 2-adrenergic receptor cytoplasmic loops. Biochemistry 44:2284–2292

Kralj JM, Douglass AD, Hochbaum DR et al (2011) Optical recording of action potentials in mammalian neurons using a microbial rhodopsin. Nat Methods 9:90–95

Kramer RH, Fortin DL, Trauner D (2009) New photochemical tools for controlling neuronal activity. Curr Opin Neurobiol 19:544–552

Kravitz AV, Freeze BS, Parker PR et al (2010) Regulation of parkinsonian motor behaviours by optogenetic control of basal ganglia circuitry. Nature 466:622–626

Lagali PS, Balya D, Awatramani GB et al (2008) Light-activated channels targeted to ON bipolar cells restore visual function in retinal degeneration. Nat Neurosci 11:667–675

Lee JH, Durand R, Gradinaru V et al (2010) Global and local fMRI signals driven by neurons defined optogenetically by type and wiring. Nature 465:788–792

Li X, Gutierrez DV, Hanson MG et al (2005) Fast noninvasive activation and inhibition of neural and network activity by vertebrate rhodopsin and green algae channelrhodopsin. Proc Natl Acad Sci USA 102:17816–17821

Lin D, Boyle MP, Dollar P et al (2011) Functional identification of an aggression locus in the mouse hypothalamus. Nature 470:221–226

Lin JY, Lin MZ, Steinbach P et al (2009) Characterization of engineered channelrhodopsin variants with improved properties and kinetics. Biophys J 96:1803–1814

Luo L, Callaway EM, Svoboda K (2008) Genetic dissection of neural circuits. Neuron 57:634–660

Madisen L, Zwingman TA, Sunkin SM et al (2010) A robust and high-throughput Cre reporting and characterization system for the whole mouse brain. Nat Neurosci 13:133–140

Nagel G, Szellas T, Huhn W et al (2003) Channelrhodopsin-2, a directly light-gated cation-selective membrane channel. Proc Natl Acad Sci USA 100:13940–13945

Oh E, Maejima T, Liu C et al (2010) Substitution of 5-HT1A receptor signaling by a light-activated G protein-coupled receptor. J Biol Chem 285:30825–30836

Osakada F, Mori T, Cetin AH et al (2011) New rabies virus variants for monitoring and manipulating activity and gene expression in defined neural circuits. Neuron 71:617–631

Rickgauer JP, Tank DW (2009) Two-photon excitation of channelrhodopsin-2 at saturation. Proc Natl Acad Sci USA 106:15025–15030

Sawadsaringkarn Y, Kimura H, Maezawa Y et al (2012) CMOS on-chip optoelectronic neural interface device with integrated light source for optogenetics. J Phys Conf Ser 352:012004

Schobert B, Lanyi JK (1982) Halorhodopsin is a light-driven chloride pump. J Biol Chem 257:10306–10313

Sohal VS, Zhang F, Yizhar O et al (2009) Parvalbumin neurons and gamma rhythms enhance cortical circuit performance. Nature 459:698–702

Stuber GD (2010) Dissecting the neural circuitry of addiction and psychiatric disease with optogenetics. Neuropsychopharmacology 35:341–342

Stuber GD, Hnasko TS, Britt JP et al (2010) Dopaminergic terminals in the nucleus accumbens but not the dorsal striatum corelease glutamate. J Neurosci 30:8229–8233

Tecuapetla F, Patel JC, Xenias H et al (2010) Glutamatergic signaling by mesolimbic dopamine neurons in the nucleus accumbens. J Neurosci 30:7105–7110

Tomita H, Sugano E, Fukazawa Y et al (2009) Visual properties of transgenic rats harboring the channelrhodopsin-2 gene regulated by the thy-1.2 promoter. PLoS One 4:e7679

Tomita H, Sugano E, Isago H et al (2010) Channelrhodopsin-2 gene transduced into retinal ganglion cells restores functional vision in genetically blind rats. Exp Eye Res 90:429–436

Tonnesen J, Sorensen AT, Deisseroth K et al (2009) Optogenetic control of epileptiform activity. Proc Natl Acad Sci USA 106:12162–12167

Tsai HC, Zhang F, Adamantidis A et al (2009) Phasic firing in dopaminergic neurons is sufficient for behavioral conditioning. Science 324:1080–1084

Wall NR, Wickersham IR, Cetin A et al (2010) Monosynaptic circuit tracing in vivo through Cre-dependent targeting and complementation of modified rabies virus. Proc Natl Acad Sci USA 107:21848–21853

Wentz CT, Bernstein JG, Monahan P et al (2011) A wirelessly powered and controlled device for optical neural control of freely-behaving animals. J Neural Eng 8:046021

Zhang F, Aravanis AM, Adamantidis A et al (2007a) Circuit-breakers: optical technologies for probing neural signals and systems. Nat Rev Neurosci 8:577–581

Zhang F, Gradinaru V, Adamantidis AR et al (2010) Optogenetic interrogation of neural circuits: technology for probing mammalian brain structures. Nat Protoc 5:439–456

Zhang F, Prigge M, Beyriere F et al (2008) Red-shifted optogenetic excitation: a tool for fast neural control derived from *Volvox carteri*. Nat Neurosci 11:631–633

Zhang F, Wang LP, Boyden ES et al (2006) Channelrhodopsin-2 and optical control of excitable cells. Nat Methods 3:785–792

Zhang F, Wang LP, Brauner M et al (2007b) Multimodal fast optical interrogation of neural circuitry. Nature 446:633–639

Zhang Y, Ivanova E, Bi A et al (2009) Ectopic expression of multiple microbial rhodopsins restores ON and OFF light responses in retinas with photoreceptor degeneration. J Neurosci 29:9186–9196

Zhao S, Cunha C, Zhang F et al (2008) Improved expression of halorhodopsin for light-induced silencing of neuronal activity. Brain Cell Biol 36:141–154

Part V
Molecular Biological Techniques

Chapter 9
Detecting Neural Activity-Dependent Immediate Early Gene Expression in the Brain

Kazuhiro Wada, Chun-Chun Chen, and Erich D. Jarvis

Abstract In this chapter, we present an in situ hybridization protocol with radioactive probe that has been successfully and easily used on detecting mRNA expression level and patterns, in multiple tissue types and developmental stages. To detect behaviorally regulated, i.e., motor or sensory, mRNA expression of immediate early genes (IEGs) within cells and tissues in vivo, in situ hybridization is a powerful method for discovering neural activity correlations and novel neural structures. Compared with nonradioactive probe methods such as digoxigenin (DIG) labeling, the radioactive probe hybridization method provides a semi-linear relation between signal intensity and targeted mRNA amounts for quantitative analysis. Furthermore, this method allows us high-throughput mRNA expression analysis for 100–200 sides with 400–1,000 tissue sections simultaneously. This method allows identifying the possible significance and function of interested genes in the nervous system under specific behaviors.

Keywords Activity-dependent gene expression • High-throughput in situ hybridization • Immediate early genes (IEGs) • Radioactive in situ hybridization • Songbird • Vocalization

9.1 Introduction

In nature, many types of behaviors are observed in a wealth of animal species at specific conditions and at specific developmental stages with species-specific features. To understand the neural basis of such behaviors, one critical step is to

---

K. Wada (✉)
Department of Biological Sciences, Hokkaido University, Sapporo, Hokkaido, Japan
e-mail: wada@sci.hokudai.ac.jp

C.-C. Chen • E.D. Jarvis
Department of Neurobiology, Howard Hughes Medical Institute, Duke University Medical Center, Durham, NC, USA

identify the specific neural circuits that are activated by production of the behavior. This first step can be achieved with activity-dependent genes, which serve as molecular markers of neural activity to map functional domains and cells of the brain. This "behavioral molecular brain-mapping approach" has been successively used to identify and characterize neural systems involved in perceiving and producing behaviors (Jarvis and Nottebohm 1997; Jarvis et al. 2000; Mello et al. 1992). Transcription of the early neural activity-induced genes, known as immediate early genes (IEGs), in the postsynaptic cell is initiated by presynaptic action potential firing and subsequent neurotransmitter and/or neuromodulator release, followed by binding to postsynaptic receptors (Clayton 2000; Worley et al. 1987). After binding, extracellular Ca^+ influx or release of Ca^+ from intracellular stores activates a signal transduction cascade, including several kinds of protein kinases, such as protein kinase A, mitogen-activated protein kinases, and calcium- and calmodulin-dependent kinases. These kinases regulate nuclear gene expression via phosphorylation of targeted specific transcription factors, including cAMP response element-binding protein (CREB)/activating transcription factor (ATF) family and serum response factors (SRFs). These transcription factors bind to promoter regions of IEGs and initiate the mRNA transcription of IEGs at several sites on chromosomes. This IEG transcription response occurs within a few minutes after neuron activation, because induction of IEG mRNAs does not require de novo protein synthesis (Guzowski et al. 1999).

The initial IEGs discovered were transcription factors that regulate other genes (Cole et al. 1989; Greenberg et al. 1986). Subsequent molecular studies discovered other types of IEGs, encoding a diverse range of functional proteins, including regulatory transcription factors, structural proteins, signal transduction proteins, growth factors, and enzymes (Loebrich and Nedivi 2009; Saffen et al. 1988; Wada et al. 2006). Thus, IEGs fall into two subcategories, the inducible transcription factors (ITFs) and all other molecules as inducible direct effectors (IDEs). There are at least two popularly studied ITFs, c-fos and egr1 (also named zif268, NGFI, Krox-24, or zenk), and one late effector, Arc [activity-regulated cytoskeleton-associated gene (Steward et al. 1998)] in vertebrates. Along with another direct effector we have studied, called dual specificity phosphatase 1 (dusp1), the combination of these four genes can be used to ensure identifying activation in all neuron types in the vertebrate forebrain (Horita et al. 2010). In invertebrates, especially in honeybees, some IEGs identified, such as *kakusei*, are induced in the mushroom body of reorienting bees and foragers (Kiya et al. 2007). Whether a transcription factor or direct effector, mRNA of most IEGs is usually induced and accumulated in the cellular cytoplasm at maximal level up to 30 min during repeated production of animal's behavior repeated sensory stimulation.

To detect behaviorally regulated (motor or sensory) mRNA expression of IEGs within cells and tissues in vivo, in situ hybridization is a powerful method. It allows single-cell resolution in tissue sections after an animal has performed behavior and processed a sensory stimulus. In situ hybridization is also a useful approach to determine the spatial and temporal profile of the IEGs of interest, especially in the absence of an available antibody. In this chapter, we present an in situ hybridization protocol with radioactive probes that has been successfully and easily used on detecting mRNA expression level and patterns, in multiple tissue types and developmental stages including embryos (Chen et al. 2012; Haesler et al. 2004).

This protocol has four major steps: (I) obtaining and sectioning the brains from behaving animals under well-controlled conditions in Sects. 9.1–9.2, (II) synthesis of radioactive RNA probes from DNA plasmids in Sects. 9.3–9.4, (III) prehybridization and hybridization using an oil bath and washing steps in Sects. 9.5–9.6, and (IV) signal detection in Sect. 9.7.

This method allows for handling of hundreds of slides simultaneously and quantitative analyses of gene expression. Compared with nonradioactive probe methods such as digoxigenin (DIG) labeling, the radioactive probe hybridization method does not require multiple amplification steps using horseradish peroxidase (HRP) antibodies and/or tyramide signal amplification (TSA) to detect signals of target probes. Therefore, this method provides a semi-linear relation between signal intensity and targeted mRNA amounts for quantitative analysis. Furthermore, compared with combination study with laser capture microdissection (LCM) and qPCR, this method allows us high-throughput mRNA expression analysis for 100–200 sides (with 400–1,000 tissue sections) simultaneously. Using these advantages, our group and colleagues have succeeded in discovering neural activity correlations and novel neural structures (Burmeister et al. 2005; Jarvis et al. 1998; Jarvis et al. 1997; Mello and Clayton 1994; Mouritsen et al. 2005) and performing high-throughput expression analysis with hundreds of genes (Wada et al. 2006).

9.2 Behavioral Observation and Brain Sampling

1. Using your behavioral paradigm of interest, keep animals in a quiet state for at least 2–3 h before performing the behavioral manipulations. Then, for at least 30 min, record the animal's behavior, such as behavioral duration, frequency, and timing, and context (with other animals, especially opposite sex) of interest {Note *1}.
2. Harvest fresh brain tissue within 5–10 min.
3. Rinse the brain tissue with 1× PBS to remove blood and feather/hair, and *carefully* remove away extra 1× PBS from the brain.
4. Embed the tissue into a pre-labeled plastic embedding mold [22×22×22 mm, PolyScience, cat# 18646A-1] and fill with OCT tissue compound [Sakura, cat#4583], avoiding air bubbles, and orientate the tissue as needed for sectioning.
5. Quickly freeze the block by placing it into crushed powder of dry ice {Note *2}.

Note

*1: Information on observation time and day, animal's developmental stage (i.e., post-birth day), housing condition, and experimental environments (new or familiar cage/room, or which field location) is also critical.

*2: This crushed dry ice powder and labeling of the plastic embedding mold should be prepared before collecting the animal brain. Do not let the animal hear extraneous noise, as this will cause IEG induction in auditory pathways. After behavior or sensory stimuli have been completed, then within the next 5–10 min, the animals should be sacrificed and the brain collected to avoid RNA degradation and handling time effect. Cooling speed should be quickened by adding ethanol as ethanol-dry ice bath.

9.3 Brain Sectioning and Stock

1. Put the OCT frozen block on sectioning stage in a cryostat and equilibrate to −20 °C for 30 min (Fig. 9.1a).
2. Trim excess OCT by a razor blade (Fig. 9.1b).
3. Slice the sample into 10–12 μm {Note *1} thick sections (Fig. 9.1c). For brain and embryonic tissue, the best cutting temperature is within the range of −18 to −20 °C {Note *2}.
4. Mount frozen sections on the face of Superfrost Plus Microscope Slides [Fisher, cat#12-550-15] {Note *3} (Fig. 9.1d–e).

Note

*1: Optimal section thickness for obtaining good-quality sections depends on tissue types, age, and animal species. We recommend performing test sectioning and Nissl staining.

*2: It is critical to adjust the cryostat temperature for the appropriate tissue type. The position of the antiroll plate should allow space between the knife and glass plate and also not touch the block when cutting. A thin, firm brush allows unfolding and other manipulation of sections.

*3: Glass slides must be chemically coated, such as with silane, to reduce occurrence of detachment of tissue slices during procedures.

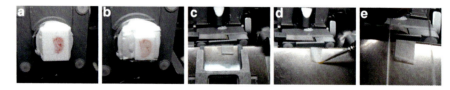

Fig. 9.1 Brain sectioning in a cryostat. (**a**) Setting of OCT frozen block on sectioning stage. (**b**) Trimmed OCT frozen block. (**c**) Slicing of the sample. (**d**) Unfolding of sliced samples using a fine brush. (**e**) Mounting of frozen section on the slide

9.4 Generation of PCR Fragments from Plasmid DNA as a Template for RNA Probe Synthesis

1. Amplification of PCR fragment that includes the gene of interest (GOI) and RNA polymerase binding sites, such as T7, T3, and Sp6 sequences, at both ends of GOI. For this purpose, we regularly use pGEM-T Easy Vector that possesses T7 and Sp6 sites, M13For and Rev sites, and restriction enzyme cloning sites (Fig. 9.2). Probe size, i.e., GOI size, can be adapted from 150 to 3,000 bp.
2. Perform PCR reaction.

    2.1 PCR content

    20 μL of distilled water (GIBCO, cat#10977)
    1 μL (~0.1 μg) of DNA template, e.g., pGEM-T easy plasmid DNA including GOI {Note *1}
    1.5 μL of M13Rev oligo primer (5′-ACAGGAAACAGCTATGACC-3′: 20 μM)
    1.5 μL of M13For oligo primer (5′-TGTAAAACGACGGCCAGT-3′: 20 μM)
    3 μL of 10× DNA polymerase buffer
    2.5 μL of 2.5 mM each dNTPs (included with Taq polymerase kit)
    0.5 μL of Taq DNA polymerase [5 unit/μL, Takara Ex taq cat#RR001]
    Subtotal 30 μL in a tube × four tubes = total 120 μL

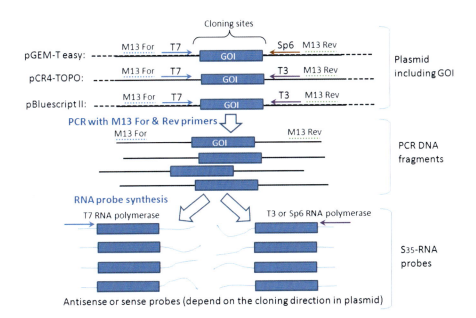

Fig. 9.2 Flow chart for making RNA probe synthesis from DNA plasmid

2.2 PCR condition

    1st cycle: 94 °C 5 min
    25–30 cycles: (94 °C 1 min +53 °C 1 min +72 °C 4 min) {Note*2}
    Last cycle: 72 °C 7 min

3. Electrophoresis on an agarose gel (1–2 %) {Note*3}.
4. Apply 15–20 μL PCR product in each well, and cut gel pieces into tubes {Note*4}.
5. Purify DNA fragments from agarose gel using a commercially available kit, such as GENECLEAN Kit [QBiogene, cat#1001-200] {Note*5}.
6. Adjust volume to 100 μL with dH$_2$O.
7. Phenol extraction and ethanol precipitation.

    7.1 Add 100 μL of PCI (phenol:chloroform:isoamyl alcohol 25:24:1) solution and vortex well.
    7.2 Centrifuge at max speed of a benchtop centrifuge machine for 3 min, and pipette off aqueous solution without disturbing or removing the pellet.
    7.3 Transfer supernatant solution into a new tube avoiding the phenol layer.
    7.4 Add 5 μL of 5 M NaCl.
    7.5 Add 250 μL of 100 % EtOH.
    7.6 Incubate the tube at −80 °C at least for 15 min.
    7.7 Centrifuge at max speed (at over 15,000 rpm) for 15 min, and get pellet.
    7.8 Rinse with 200 μL of 70 % EtOH.
    7.9 Centrifuge at max speed of a benchtop centrifuge machine for 3 min, and pipette off aqueous solution without disturbing or removing the pellet.
    7.10 Add 20 μL RNase-free water, such as Pure Water (Invitrogen, cat#10977-015).

8. Use 1 μL for checking of DNA concentration on an agarose gel.
9. Take another 1 μL for DNA spectrometer, and adjust DNA concentration to 0.25 μg/μL {Note*6}.
10. Store the tube in a −20 °C freezer {Note*7}.

Note

*1: Do not include more than 1 μg plasmid DNA. Nonspecific bands may be amplified.
*2: PCR cycles should be <30 to prevent amplifying nonspecific bands.
*3: Use thick and clean gels and fresh TAE buffer in the electrophoresis equipment.
*4: It is important to cut gel pieces including PCR bands on an UV illuminator as quickly to avoid DNA nicks by UV light.
*5: An alternative method to gel purification is the QIAquick PCR purification kit from Qiagen.
*6: If the concentration of the DNA solution is low, adjust to 0.25 μg/μL by evaporation or EtOH precipitation.
*7: It is possible to store the purified DNA solution for more than 3–4 years at −20 °C.

9.5 S³⁵-RNA Probe Synthesis

1. Use 0.5 mL tube or 0.2 mL PCR tube.
2. Add and mix contents below.

   1.0 µL of purified PCR DNA fragment (0.25 µg/µL)
   1.0 µL of 10× Roche reaction buffer
   0.3 µL of RNasin [40 U/µL: Promega, cat#N251B]
   1.5 µL of 10 mM AGC mix solution {*Note *1*}
   4.5 µL of S³⁵-UTP [PerkinElmer, cat#NEG-039H] {*Note *2*}
   0.7 µL of pure water
   1.0 µL of T7, T3 or Sp6 RNA polymerase [20 unit/µL: Roche, cat#10881767001, 1031171001, 11487671001, respectively]

3. Incubate the reaction tube immediately at 37 °C for 2 h {*Note *3*}.
4. Fill up to 50 µL with 40 µL of pure water.
5. Add 2.5 µL of 5 M NaCl.
6. Add 125 µL of 100 % EtOH, and then mix very well.
7. Incubate the tube at −80 °C or on dry ice at least for 15 min.
8. Centrifuge at max. speed at 4 °C for more than 15 min, and discard supernatant.
9. Wash pellet with 300 µL of 70 % EtOH and centrifuge again at max speed at 4 °C, and discard supernatant.
10. Dissolve pellet with 10 µL of pure water by pipetting.
11. Add 40 µL of hybridization solution {*Note *4*} and mix well (then store the solution at −20 °C).
12. For checking radioactive counts, pipette 1 µL of S³⁵-RNA probe solution into 1 mL scintillation cocktail in a plastic counting vial, and mix very well.

    (Add 10⁶ cpm of S³⁵ cRNA probe solution for 100 µL hybridization solution.)

    Note

    *1: Preparation of 10 mM AGC mix solution.
    Mix 2 µL of 100 mM ATP [Roche, cat#1140965], 2 µL of 100 mM GTP [Roche, cat#1140957], 2 µL of 100 mM CTP [Roche, cat#1140922], and 54 µL of pure water.
    Stock mixed solution at −20 °C.

    *2: Half-life of S³⁵ radioactivity is 87.51 days. Therefore, it is better to use shipped S³⁵-UTP solution within 1–1.5 months.

    *3: To prohibit condensation formed on top of a tube that affects reaction efficiency, use a PCR machine with heat cover.

*4: Preparation of in situ hybridization solution.

For total 10 mL volume:
Mix 5 mL of 100 % formamide + 600 μL of 5 M NaCl + 100 μL of 1 M Tris–HCl pH 8.0 + 240 μL of 0.5 M EDTA pH 8.0 + 200 μL of 50× Denhart's solution + 100 μL of 1 M DTT + 250 μL of 20 mg/mL tRNA [Roche, cat#109495] + 1 g of sodium dextran sulfate 500,000.

Finally, bring volume up to 10 mL with pure water.

Much time is needed to dissolve sodium dextran sulfate by shaking at room temperature.

Stock the solution in −20 °C.

9.6 Pre-hybridization and Hybridization

1. Prepare appropriate amounts of 4 % paraformaldehyde/1× PBS {Note*1}, 1× PBS {Note*2}, and 2× SSPE {Note*3}, and set each container (Fig. 9.3).
2. Immerse glass slides with tissue sections in 4 % paraformaldehyde/1× PBS for 5 min at room temperature (RT) {Note*4}.
3. Rinse three times in 1× PBS in three separate containers, 2 min each, with occasionally gentle shaking.
4. Put slides in 1 L of acetylation solution {Note*5} for 10 min.
5. Rinse three times in 2× SSPE for each 2 min in three separate containers.
6. Dehydrate through the alcohol series, 50 % EtOH, 70 % EtOH, 95 % EtOH, and then 100 % EtOH {Note*6}, for 2 min each.
7. Let the slides dry under the hood on a paper towel.
8. Calculate the total volume needed for all slides (plus a few extra), from 50 to 150 μl {Note*7} of hybridization solution per slide.
9. Mix S^{35}-cRNA probe and hybridization solution very well, but gently, avoiding generation of air bubbles (adjust 10^6 cpm S^{35}-cRNA probe/100 μL hybridization solution).

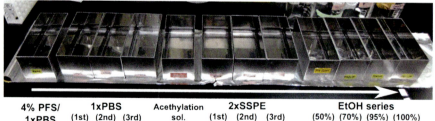

Fig. 9.3 Solution series for pre-hybridization

9 Detecting Neural Activity-Dependent Immediate Early Gene Expression... 141

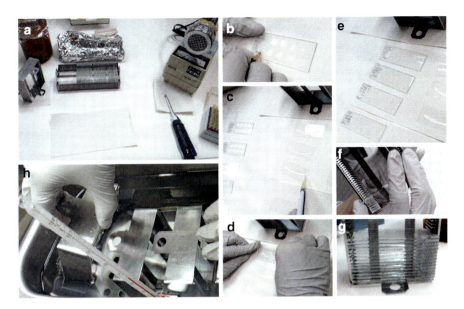

Fig. 9.4 Cover slip procedure and hybridization in oil bath with radioactive RNA probe. (**a**) Setting of cover slip procedure. (**b**) Labeling sample information on glasses with a pencil. (**c**) Putting S35-cRNA probe/hybridization solution on a cover slip. (**d** & **e**) Placing a glass slide on a cover slip, then immediately flip the side to face up. (**f** & **g**) Placing the cover-slipped slides in a metal rack. (**h**) Placing the metal racks into the oil bath

10. Incubate mixed solution at 65 °C water bath for 5 min to dissolve reagents, and then immediately chill on ice at least for 5 min.
11. Put S^{35}-cRNA probe/hybridization solution on a cover slip, and gently place a glass slide on it with the tissue sections facing the hybridization solution (Fig. 9.4a–e).
12. Place the cover-slipped slide in a metal rack, making sure the slides stay in a horizontal position (Fig. 9.4f–g).
13. Carefully place each rack into the oil bath *{Note*8}* at 65 °C *{Note*9}* (Fig. 9.4h).
14. Incubate slides overnight (12–15 h) *{Note*10}*.

    Note

    *1: Preparation of 3 % paraformaldehyde/1× PBS:

    For 1 L solution:

    Mix 30 g paraformaldehyde in 100 mL of 10× PBS + 900 mL deionized H_2O (dH_2O, no need to use DEPC-H_2O), and add 320 µL of 10 N NaOH (to help paraformaldehyde to dissolve, adjustment of pH).

    While stirring, heat the solution (~40 °C) under a hood on the hot plate.

    *2: Preparation of 1× PBS:

    ten times dilution from 10× PBS.

To make 10× PBS for 1 L, add 800 mL of dH$_2$O to a large beaker.
While the water is stirring with a magnet on stir plate, add the following:
80 g NaCl
2 g KCl
29 g Na$_2$HPO$_4$-12H$_2$O
2 g KH$_2$PO$_4$
Finally, fill up to 1 L.

*3: Preparation of 2× SSPE:
Ten times dilution from 20× SSPE. For making 5 L of 20× SSPE solution, mix 876.5 g of NaCl, 153.5 g of NaH$_2$PO$_4$-2H$_2$O, and 37 g of EDTA, add 32.5 mL of 10 N NaOH, and fill up to 5 L of dH$_2$O (no need to use DEPC-H$_2$O).
Store at RT.

*4: The incubation time in 4 % paraformaldehyde/1× PBS is critical for fixation, a condition that affects hybridization efficiency later.

*5: Preparation of 1 L of acetylation solution:
Mix 13.6 mL triethanolamine in 1L dH$_2$O (no need to use DEPC-H$_2$O).
Right before you're ready to use it, add 2.52 mL of acetic anhydride, mix well, and immediately pour the solution over the slides.

*6: These alcohol solutions can be reused at least 5–10 times, when processing ~60 slides each time.

*7: The amount of hybridization solution depends on how many tissue sections are attached on the glass slide.

*8: Mineral oil [Sigma, cat#330760] is used.

*9: Hybridization temperature: We recommend 65 °C as a default of hybridization and wash temperature. If high background signal was observed as a result of an RNA probe size (over 3 kbp) or high G/C % (over 70 %), it would be good to try 70 °C. Conversely, if a researcher wants to try cross-species hybridization with RNA probes generated from a different species' cDNA sequence and does not obtain a strong enough, high-quality signal, then try lower hybridization and wash temperatures in steps of 5 °C (However, in such case, cloning and using the species-specific cDNA fragment of GOI is the better approach.).

*10: Hybridization time is one of the most crucial points to affect signal intensity and noise ratio.

We recommend using a standard hybridization time (~15 h) across experiments, if a researcher wants to compare the results from experiments performed on different days (but using the same S^{35} RNA probes). Over 20 h incubation usually generates high background. Although general signal intensity may be less, 6–9 h of incubation may generate a better S/N ratio.

9.7 Post-hybridization Washing Treatment

1. Prepare solutions and set each container (Fig. 9.5).
2. To remove mineral oil, immerse the metal rack into at least two consecutive chloroform washes. Between transfers of chloroform washes, dip the solution off the rack very well.
3. Place the metal rack with slides into 2× SSPE+0.1 % β-mercaptoethanol.
4. Remove the cover slips from the slides with forceps while the slides are in this solution and quickly.
5. Place slides into a metal rack in 2× SSPE+0.1 % β-mercaptoethanol without letting the section dry {Note *1}.
6. Incubate slides in 2× SSPE+0.1 % β-mercaptoethanol for 30 min at RT with occasional shaking {Note *1}.
7. Incubate slides in pre-warmed 50 % formamide/2× SSPE/0.1 % β-mercaptoethanol for 1 h at 65 °C with occasional (two to three times) shaking {Note *1}.
8. Incubate in pre-warmed 0.1× SSPE/0.1 % β-mercaptoethanol for 30 min at 65 °C with occasional shaking {Note *1}.
9. Repeat the same procedure (0.1× SSPE/0.1 % β-mercaptoethanol for 30 min at 65 °C) again in another new solution.
10. Run through the alcohol series, 50 % EtOH, 70 % EtOH, 95 % EtOH, and 100 % EtOH, for 2 min each {Note *2}.
11. Let the slides dry under the hood on a paper towel.

Note

*1: After rinsing, this 2× SSPE+0.1 % β-mercaptoethanol must be handled as radioactive waste.
*2: These alcohol solutions can be reused over 10 times, with ~60 slides per time.

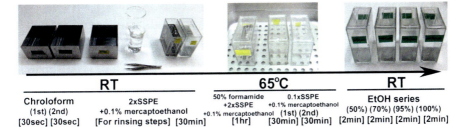

Fig. 9.5 Solution series for post-hybridization washing

9.8 Signal Detection of S[35]-RNA Probe: Visualization of Radioactive Signal

Detection of Signal with X-ray Film:

1. Place dry glass slides into a film cassette and expose the slides to X-ray film (Kodak, BioMax MR film) in a dark room for several days {*Note*1*} (Fig. 9.6a). Make sure that the slides face the emulsion side of the X-ray film {*Note*2*}.
2. Develop the X-ray film in standard developer and fixer in a dark room {*Note*3*} (Fig. 9.6b).
3. The hybridization signal shows up as black (exposed silver grains in the emulsion) on the film.

Note

*1: For test exposure, usually 1–2 days is enough. For regular exposure, 2–4 days should be fine.

*2: We do not recommend using regular X-ray film that has emulsion on both sides, because S[35] radioactivity cannot go through opposite side of the X-ray film. Therefore, regular X-ray films cannot be enhanced with S[35] in situ hybridization; rather, they cause a more diffused image.

*3: If needed to perform manual development of X-ray film, prepare one developer container, two tap water containers, and one fixer container in a dark room.

Under safelight conditions, put X-ray films in a developer container for 3 min, then transfer it to the first tap water container for 1 min to rinse out developer solution. Immerse films in a fixer container for 3 min, transfer it to the second tap water container for a few minutes rinsing very well, and then hang dry them (the later step can be at regular room light).

(Optional)

Detection of Signal with High Resolution by Silver Grain Dipping:

For Generating of Higher Resolution Signals with Cresyl Violet {*Note *1*}:

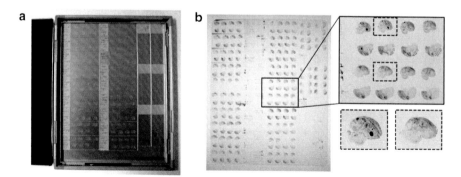

Fig. 9.6 Visualization of radioactive signal with X-ray film. (**a**) Setting of hybridized glasses on a film casette. (**b**) Developed X-ray film. Black color represents mRNA signals

1. Delipidize slides by incubating them in xylene (rack needs to be mettle) for 5 min at RT twice {Note *1}.
2. Remove xylene and rehydrate 1 min each in 100 %, 100 %, 95 %, 95 %, 70 %, and 50 % EtOH and then in dH$_2$O {Note *1}.
3. Dry slides well under a hood for at least 2–3 h.
4. Then, in the dark room under red-color safelight, scoop out needed amount of Kodak NTB2 emulsion into 50 mL tube with the same amount of dH$_2$O to make 1:1 ratio solution.
5. Incubate the emulsion solution in a 42°C water bath at least 15 min, occasionally gently mixing but not to make air bubbles.
6. Dip glass slides into the diluted emulsion in the 42°C water bath.
7. Dry dipped slides in a closed lighttight container with silica gel overnight in an oven at 37 °C.
8. Transfer the slides into the black boxes containing silica gel {Note *2}.
9. Seal the edges of the boxes with black electrical tape (avoiding generating static electricity) and wrap the boxes in aluminum foil.
10. Store the boxes at 4 °C {Note *3} from several days to weeks {Note *4}.

For Developing Silver Dipping Slides:

11. Warm up the slide boxes to room temperature for 1 h.
12. In the darkroom, set five metal trays with two developer [Kodak, cat#1464593], one tap water, and two fixers [Kodak, cat#1971746] and a large plastic tray for final rinsing with tap water.
13. Develop the dipped slides in Kodak D-19 developer at 16 °C for 3 min, twice.
14. Wash the developed slides in tap water at RT for 1 min.
15. Incubate the slides in fixer at 19 °C for 3 min, twice.
16. Then, lights can be turned on.
17. Wash the slides in running tap water at RT at least 10 min.
18. While washing, with a razor blade scrape the emulsion off the backside of the slide that does not contain tissue sections {Note *5}, and rinse with tap water.
19. Dry slides up slides at RT.

For Staining Tissues with Cresyl Violet:

20. Stain tissue with 0.3 % cresyl violet in tap water solution for 15 min in a 37 °C water bath {Note *6}.
21. Wash excess cresyl violet solution in fresh tap water for ~15 dips.
22. Dehydrate the slides for ~15 dips each in alcohol solutions: 50 %, 70 %, 95 %, 95 %, 100 %, and 100 % EtOH.
23. Incubate the slides in xylene (rack needs to be metal) for 5 min at room temperature twice.
24. Place Permount medium [Fisher, cat#SP15-500] on the slide, cover slip with a glass cover slip, and dry the covered slide in the hood for overnight.
25. The sections can be examined underneath the microscope within 1 day, but it takes about 72 h for the Permount to become hard enough to clean the slide without the cover slip coming off.

Note

*1: If not going to stain with cresyl violet, then delipidization is not necessary. It is necessary to delipidize before placing slides in emulsion as the lipids will not be removed easily after covering with emulsion.
*2: At this step, make sure the condition of the dipped slides is completely dry.
*3: Keeping dipped slides at 4 °C is critical to avoid mold growing on the glass slides during exposure of the S^{35} signal to the emulsion.
*4: Signal from 1 day on X-ray film is similar to 5 days under emulsion.
*5: The slide needs to be wet; otherwise the razor blade will scratch the slides.
*6: This step can be performed at RT. In the case, the incubation time should be longer.

9.9 Concluding Remarks

For visualizing radioactive probe signal on brain sections, there are major two ways, detection by X-ray films and by silver grains.

Detection by X-ray films provides a relatively quick result and analyses. Furthermore, the X-ray film data could be used for high-throughput quantitative analysis. The X-ray film also reveals broad anatomical resolution. It provides the ability to easily compare expression patterns among different brain sections on the same and different slides. For this purpose, it is very critical to pay close attention to obtaining a high signal-to-background noise ratio on the X-ray film. High background on X-ray films is usually caused by using old developer or fixer or with probe problem. Poor quality of riboprobes also generates high background. In our experience, probes generated with Sp6 RNA polymerase more than T7 and T3 RNA polymerases have a tendency toward such problems. In such case, we recommend to increase enzyme units for RNA synthesis. The oil creates an instant seal, allowing for rapid processing of many slides. Sometimes, mineral oil from the hybridization step remains on the slides and tissue sections, which will cause spotted dark background. For quantitative analysis, we recommend not to use overexposed films. It is critical to maintain a signal intensity that correlates linearly with actual mRNA expression levels. To show digital photo images of brain sections from X-ray film, there are two representative ways: regular bright-field image (in this case, signal is black in color: Fig. 9.7a top low) and inverted black/white image (in this case, signal is white color: Fig. 9.7a 2nd low). Inverted black/white images can be easily produced from regular bright-field images using "inverted color function" in an image software, such as Image J or Photoshop.

Detection by silver grains in the emulsion is produced by the pattern of decay emission from the radioactive probe. The emulsion results show tissue morphology and higher cellular resolution of gene expression. The emulsion contains light-sensitive silver salts placed over the tissue. After exposure and developing the emulsion, exposed silver salts are converted to metallic silver grains. The metallic silver grains block direct light through and appear as the black dots under bright-field view.

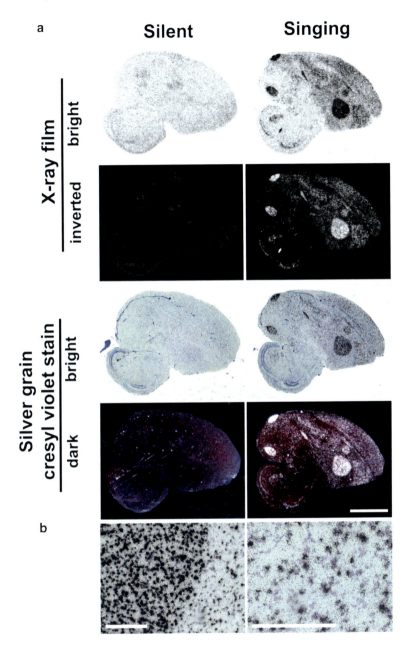

Fig. 9.7 Photo images of radioactive signals. (a) whole brain in-situ hybridization images under silent and singing conditions, using X-ray film images under bright light (top panels; black = mRNA signal), inverted X-ray film images (2nd low; white = mRNA signals), silver grain dipped-cresyl violet stained images under regular bright light (3rd low; black = mRNA signals, blue/purple = cresyl violet stained cells), and silver grain dipped-cresyl violet stained images under polarized light in dark field (4th low; white = mRNA signals, purple/red = cresyl violet stained cells). Scale bar = 2mm. (b) Higher magnification of silver grain dipped-cresyl violet stained images under regular bright light. Developed silver grains are observed as black dots. Scale bars = 200μm

In this situation, cells lable blue/purple in color after cresyl violet staining (Fig. 9.7a, 3rd low & b). The silver deposits over the cells represent mRNA gene expression and can be observed and measured qualitatively under a microscope. Gene expression can be quantified at the cellular level by counting the average number of silver grains over cells relative to the background silver grains elsewhere on the tissue or slide.

In dark field the silver grains reflect light coming from the side and appear as the white dots (Fig. 9.7a, 4th low). The cresyl violet stain appears as a purple/red color. In the dark field view, the hybridization signal is easier to visualize under lower magnification, and its image is commonly used to show the overall gene expression pattern.

References

Burmeister SS, Jarvis ED, Fernald RD (2005) Rapid behavioral and genomic responses to social opportunity. PLoS Biol 3:e363

Chen CC, Wada K, Jarvis ED (2012) Radioactive in situ hybridization for detecting diverse gene expression patterns in tissue. JVis Exp pii:3764

Clayton DF (2000) The genomic action potential. Neurobiol Learn Mem 74:185–216

Cole AJ, Saffen DW, Baraban JM, Worley PF (1989) Rapid increase of an immediate early gene messenger RNA in hippocampal neurons by synaptic NMDA receptor activation. Nature 340:474–476

Greenberg ME, Ziff EB, Greene LA (1986) Stimulation of neuronal acetylcholine receptors induces rapid gene transcription. Science 234:80–83

Guzowski JF, McNaughton BL, Barnes CA, Worley PF (1999) Environment-specific expression of the immediate-early gene Arc in hippocampal neuronal ensembles. Nat Neurosci 2:1120–1124

Haesler S, Wada K, Nshdejan A, Morrisey EE, Lints T, Jarvis ED, Scharff C (2004) FoxP2 expression in avian vocal learners and non-learners. J Neurosci 24:3164–3175

Horita H, Wada K, Rivas MV, Hara E, Jarvis ED (2010) The dusp1 immediate early gene is regulated by natural stimuli predominantly in sensory input neurons. J Comp Neurol 518:2873–2901

Jarvis ED, Nottebohm F (1997) Motor-driven gene expression. Proc Natl Acad Sci USA 94:4097–4102

Jarvis ED, Ribeiro S, da Silva ML, Ventura D, Vielliard J, Mello CV (2000) Behaviourally driven gene expression reveals song nuclei in hummingbird brain. Nature 406:628–632

Jarvis ED, Scharff C, Grossman MR, Ramos JA, Nottebohm F (1998) For whom the bird sings: context-dependent gene expression. Neuron 21:775–788

Jarvis ED, Schwabl H, Ribeiro S, Mello CV (1997) Brain gene regulation by territorial singing behavior in freely ranging songbirds. Neuroreport 8:2073–2077

Kiya T, Kunieda T, Kubo T (2007) Increased neural activity of a mushroom body neuron subtype in the brains of forager honeybees. PLoS One 2:e371

Loebrich S, Nedivi E (2009) The function of activity-regulated genes in the nervous system. Physiol Rev 89:1079–1103

Mello CV, Clayton DF (1994) Song-induced ZENK gene expression in auditory pathways of songbird brain and its relation to the song control system. J Neurosci 14:6652–6666

Mello CV, Vicario DS, Clayton DF (1992) Song presentation induces gene expression in the songbird forebrain. Proc Natl Acad Sci USA 89:6818–6822

Mouritsen H, Feenders G, Liedvogel M, Wada K, Jarvis ED (2005) Night-vision brain area in migratory songbirds. Proc Natl Acad Sci USA 102:8339–8344

Saffen DW, Cole AJ, Worley PF, Christy BA, Ryder K, Baraban JM (1988) Convulsant-induced increase in transcription factor messenger RNAs in rat brain. Proc Natl Acad Sci USA 85:7795–7799

Steward O, Wallace CS, Lyford GL, Worley PF (1998) Synaptic activation causes the mRNA for the IEG Arc to localize selectively near activated postsynaptic sites on dendrites. Neuron 21:741–751

Wada K, Howard JT, McConnell P, Whitney O, Lints T, Rivas MV, Horita H, Patterson MA, White SA, Scharff C et al (2006) A molecular neuroethological approach for identifying and characterizing a cascade of behaviorally regulated genes. Proc Natl Acad Sci USA 103:15212–15217

Worley PF, Baraban JM, Snyder SH (1987) Beyond receptors: multiple second-messenger systems in brain. Ann Neurol 21:217–229

Chapter 10
Epigenetic Regulation of Gene Expression in the Nervous System

Dai Hatakeyama, Sascha Tierling, Takashi Kuzuhara, and Uli Müller

Abstract The term "epigenetics" refers to heritable alterations in chromatin structure due to modifications of genomic DNA and histone proteins. Basic insights about epigenetic alterations are derived from investigations of cell division and development. Recently, many neurobiologists have focused on the mechanisms of epigenetic control to link gene expression with behavioral changes in animals because the long-lasting composition of epigenetic modifications is consistent with the characteristics of long-term memories. There are several kinds of epigenetic modifications: (1) cytosine methylation of genomic DNA, (2) acetylation, (3) methylation, and (4) phosphorylation of histones. In this chapter, we reviewed the fundamental techniques for investigating epigenetic status with specific focus on cytosine methylation of genomic DNA. In addition, methods for analyzing histone modifications are also briefly described.

Keywords Bisulfite PCR • Cytosine methylation • DNMT • Epigenetic • HAT • HDAC • Histone acetylation • Honeybee • MBD • Single nucleotide primer extension (SNuPE)

D. Hatakeyama (✉) • U. Müller
Zoologie/Physiologie, FR 8.3 – Bioscience, Saarland University,
Postfach 151150, 66041 Saarbrücken, Germany
e-mail: daihatake926@ph.bunri-u.ac.jp; uli.mueller@mx.uni-saarland.de

S. Tierling
Genetik/Epigenetik, FR 8.3 – Bioscience, Saarland University,
Postfach 151150, 66041 Saarbrücken, Germany
e-mail: s.tierling@mx.uni-saarland.de

T. Kuzuhara
Laboratory of Biochemistry, Faculty of Pharmaceutical Sciences,
Tokushima Bunri University, 180 Nishihama-Houji, Yamashiro-cho,
Tokushima City, Tokushima 770-8514, Japan
e-mail: kuzuhara@ph.bunri-u.ac.jp

10.1 Introduction

A variety of epigenetic modifications modulate eukaryotic genomic information affecting transcription profiles and DNA replication and repair processes. Many investigators in the field of epigenomics focus on DNA methylation and histone modifications and how these modulations interact to form stable epigenetic chromatin conformation. In this review, we focus on the methods used to assay DNA methylation.

DNA methylation involves the addition of a methyl moiety to the cytosine-5 position (Fig. 10.1) in most eukaryotic organisms. The enzymes that specifically methylate the C5 carbon of cytosines in DNA to produce C5-methylcytosine are DNA methyltransferases (DNMTs). Enzymatically, they transfer a methyl group which is derived from *S*-adenosyl-L-methionine to the C5 carbon of the pyrimidine ring. This results in the generation of 5-methyl-cytosine and *S-adenosyl-homocysteine*. Five different enzymes involved in DNA methyl-group transfer are known: DNMT1 maintains DNA methylation after replication by methylating the newly synthesized strand (Araujo et al. 1999), DNMT3A and DNMT3B methylate genomic regions de novo (Takeshima et al. 2006), DNMT2 has a tRNA methylating

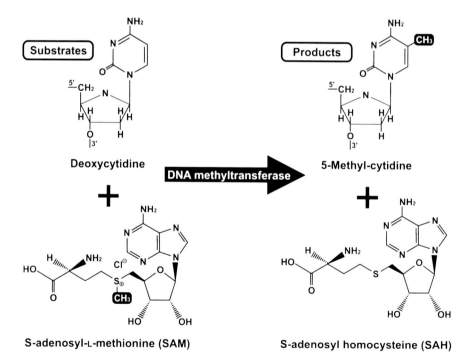

Fig. 10.1 Chemical reaction for methylation on cytosine residues

function only (Goll et al. 2006; Jurkowski et al. 2008), and DNMT3L protein lacks a catalytic domain; however, it acts as a positive regulator of DNMT3A (Chedin et al. 2002). This epigenetic modification mainly occurs in a simple dinucleotide cite, CpG. Depending on the organism, roughly 60–90 % of all CpG dinucleotides in mammals are methylated. Most methylated CpGs are scattered throughout the genome preventing transcription or transposition of retroviral elements. However, some are clustered in regions of relatively high CpG density known as "CpG islands." These are usually between 200 and 500 bp in size, often cover the promoter region of genes, and are kept methylation free. Locally, methylation of cytosine residues is the most common epigenetic modification of mammalian genomic DNA and is primarily associated with transcriptional regulation of tissue-specifically expressed genes or silencing of tumor suppressor genes in cancer cells (Novak et al. 2006), stem cells (Rodic et al. 2005) and glial cells (Hoffman and Hu 2006). Methylated cytosine residues are not only found in vertebrates but also in invertebrates, such as chordates (Simmen et al. 1999; Suzuki et al. 2007), echinoderms (Tweedie et al. 1997), and insects (Field et al. 2004; Glastad et al. 2011; Lyko and Maleszka 2011).

Especially in insects, *Drosophila melanogaster* is the most studied invertebrate species in the field of DNA methylation. *Drosophila* DNA methyltransferase (dDNMT2), which is closely related to the mammalian DNMT2 family and is encoded by a single DNA methyltransferase gene in the *Drosophila* genome (Tweedie et al. 1999; Hung et al. 1999; Tang et al. 2003; Marhold et al. 2004), putatively methylated the components of retrotransposons and repetitive DNA sequences (Salzberg et al. 2004). Overexpression of dDNMT2 resulted in significant genomic hypermethylation at CpT and CpA dinucleotides, but not CpG dinucleotides (Kunert et al. 2003). In addition, the *Drosophila* genome contains a single gene that encodes a methyl-CpG-binding domain protein (MBD2/3) with extensive homology to the vertebrate methyl-DNA binding proteins MBD2 and MBD3 (Tweedie et al. 1999; Roder et al. 2000; Ballester et al. 2001). Band shift assays have demonstrated the specific binding of MBD2/3 to CpT/A-methylated DNA (Marhold et al. 2004), which reflects the endogenous DNA methylation pattern in *Drosophila* (Lyko et al. 2000).

10.2 Honeybee as a Model Organism for Epigenetic Analyses

In this review, we focused on the honeybee *Apis mellifera* as a model organism for epigenetic analyses. The honeybee is an excellent model organism for understanding the relationship between behavioral changes and epigenetic modifications for several reasons. First, the entire genome sequencing of honeybee has been completed, which is essential for the comprehensive understanding of epigenetic modification systems (The Honeybee Genome Sequencing Consortium 2006;

Schaefer and Lyko 2007; Foret et al. 2009; Lyko et al. 2010; Gabor Miklos and Maleszka 2011). Second, extensive physiological, biochemical, and molecular biological data are available to understand the mechanisms of honeybee development and complex behaviors, such as caste system and long-term memory (Müller and Hildebrandt 2002; Park et al. 2003; Locatelli et al. 2005; Okada et al. 2007; Watanabe et al. 2006; Kamakura 2011). Finally, sufficient quantity of genomic DNA can be purified from a single honeybee brain to investigate epigenetic changes, which advantageously allows behavioral changes to be linked to epigenetic modulation at the level of individual organisms.

Recently, increasing evidence is available for understanding the molecular mechanisms of epigenetic control in honeybees. While *Drosophila* only has a single type of putative DNA methyltransferase (DNMT) (Tweedie et al. 1999; Marhold et al. 2004), honeybees possess the entire components of functionally active vertebrate-like DNMTs (Wang et al. 2006) and several isoforms of methyl-CpG-binding domain proteins (MBDs) (Wang et al. 2006). Additionally, non-CpG methylation, which is observed in the *Drosophila* genome, is either extremely rare or nonexistent in the honeybee genome (Wang et al. 2006). These data suggest that honeybees have an active DNA methylation system similar to that found in mammals. Methylated cytosine residues have been found in several genes in honeybees (Wang et al. 2006; Ikeda et al. 2011; Shi et al. 2011), and DNA methylation was shown to play crucial roles in development and caste differentiation (Kucharski et al. 2008; Elango et al. 2009; Kim et al. 2009). In this review, we focus on cytosine methylation of genomic DNA and describe fundamental techniques for investigating epigenetic status.

10.3 Dissection of Honeybee Brains and Purification of Genomic DNA

Honeybees were caught, briefly anesthetized on ice, and decapitated. Honeybee heads were embedded and fixed on the surface of melted dental wax (melting point = ~60 °C), and their brains were isolated. The brains were frozen in liquid nitrogen immediately after dissection and stored at −80 °C until used for genomic DNA purification.

Honeybee brains were homogenized in liquid nitrogen and incubated at 55 °C for 2 h in 500 µg/mL proteinase K solution diluted with DNA extraction buffer containing 50 mM Tris–HCl (pH 7.5), 100 mM NaCl, 1 mM EDTA (pH 8.0), and 1.5 % SDS. After gentle treatment twice with phenol/chloroform and once with chloroform/isoamyl alcohol, genomic DNA was precipitated with ethanol and treated with 10 µg/mL RNase A at 37 °C for 1 h. Genomic DNA was treated again with phenol/chloroform and chloroform/isoamyl alcohol, precipitated with ethanol, air-dried, dissolved in 10 mM Tris–HCl (pH 8.5), and stored at −80 °C until use. This procedure can be applied for the purification of genomic DNA from any tissue or cell type irrespective of the animal species.

10.4 Direct Detection of DNA Methylation by "Southern-Western" Blotting

We estimated and compared the methylation levels in genomic DNA from mouse, *Drosophila*, and honeybee brains by "Southern-Western" blotting. Using anti-5-methylcytosine mouse antibody easily enabled to combine Southern blotting and Western blotting.

10.4.1 Procedure for Southern-Western Blotting

To compare the relative amount of whole-genome methylation, 5 μg of honeybee, mouse, and *Drosophila* genomic DNA were run on a 0.8 % agarose gel and transferred onto a nitrocellulose membrane following a standard method for Southern blotting. After treatment with a blocking buffer containing 1 % BSA and 1 % Tween-20 in phosphate-buffered saline (PBS) for 1 h at room temperature, the membrane was incubated with anti-5-methylcytosine mouse antibody (Calbiochem) diluted to 1:1,000 with blocking buffer and incubated overnight at 4 °C. Other blocking buffers, such as 5 % skim milk and 1 % Tween-20 in PBS, could have also been used for this experiment. The membrane was rinsed with washing buffer (0.1 % Tween-20 in PBS; 3 × 15 min) and incubated with horseradish peroxidase (HRP)-labeled anti-mouse IgG diluted to 1:10,000 with blocking buffer at room temperature for 1 h. The membrane was then rinsed with washing buffer (3 × 15 min) and treated with chemiluminescent HRP substrate for 5 min at room temperature.

10.4.2 Results of Southern-Western Blotting

Although equal amounts of genomic DNA (5 μg) purified from honeybee, *Drosophila*, and mouse brains were subjected to agarose gel electrophoresis (Fig. 10.2a), the intensity of the band from the honeybee genomic DNA was lower than that from mouse and *Drosophila* genomic DNA (Fig. 10.2b). This data supports a previous report that found a smaller amount of methylcytosine in honeybee and *Drosophila* genomic DNA than mammalian DNA (Wang et al. 2006).

10.5 Methylation-Sensitive Restriction

Besides the enrichment of methylated DNA by immunoprecipitation (which is pretty much dependent of local CpG content and DNA accessibility), methylation-specifically cutting endonucleases can be used for investigating methylation

Fig. 10.2 Immunoblotting with anti-5-methylcytosine antibody. (**a**) Staining of the agarose gel with ethidium bromide showed that the amount of genomic DNA loaded from mouse, honeybee, and *Drosophila* brains was similar. (**b**) Although the honeybee genomic DNA also had methylated cytosines, the amount in the total genomic DNA was significantly less compared to the mouse and *Drosophila* genomes

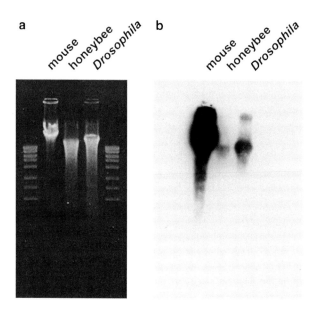

patterns in genomic DNA. Enzymes such as *Hpa* II or *Hha* I only cut unmethylated DNA leaving methylated DNA untouched. Subsequent Southern blotting provides information about the relatively wide occurrence and abundance of 5-methylcytosine in the genome (Rae and Steele 1979; Quint and Cedar 1981). This is a classical method for investigating methylation patterns and recently used for invertebrates (Krauss et al. 2009; Robinson et al. 2011). However, both immunoprecipitation and methylation-sensitive restriction require a large amount of purified genomic DNA which is, especially when using tiny amounts of biological material, not always available.

10.6 Bisulfite PCR

Since more than a decade, bisulfite conversion was developed for the sensitive identification and direct mapping of the site of 5-methylcytosine using a very small quantity of genomic DNA (Frommer et al. 1992). This technique breaks down epigenetic information to the genetic level and is widely used for local, and nowadays also genome-wide, methylation detection. The standard method for this technique is based on the chemical conversion of unmethylated cytosines to uracils by treatment of DNA with sodium bisulfite (Fig. 10.3). This chemical modification precedes in three steps: (1) sulfonation at the C6 position of the cytosine residue, (2) hydrolytic deamination at the C4 position to produce uracil sulfonate, and (3) desulfonation under alkaline conditions. The 5-methylcytosine remains unreactive to this process

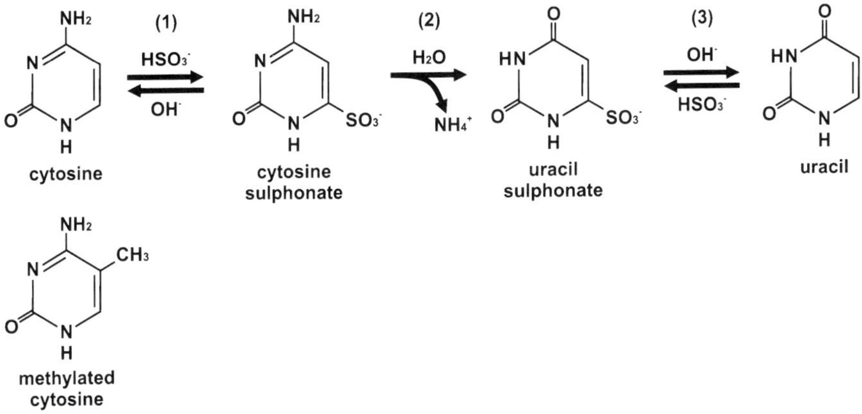

10 Epigenetic Regulation of Gene Expression in the Nervous System

Fig. 10.3 Chemical reaction for the bisulfite conversion of cytosine to uracil

since hydrolytic deamination caused by bisulfite at the C6 position is blocked by the presence of the methyl group at C5. Using next-generation-sequencing techniques like the HiSeq2000 platform (Illumina), high-resolution genome-wide DNA methylation patterns with high coverage can be obtained. For local profiling, bisulfite-specific PCRs can be performed using a strand-specific primer set detecting the methylation information on one strand of the DNA. During PCR, 5-methylcytosine residues are amplified as cytosines and unmethylated cytosines are amplified as thymines (Fig. 10.4a). The PCR products are cloned into proper plasmids and sequenced using Sanger chain termination technology (Sanger et al. 1977). Comparison of the sequence of bisulfite-treated DNA to genomic DNA allows the identification of 5-methylcytosine sites. To standardize the identification of methylated cytosines within the amplicon, semiautomated tools are available (BiQ Analyzer, Bock et al. 2005). For local profiling, next-generation-sequencing techniques are also available by now using the 454 GS-FLX pyrosequencing platform (Roche) with a freely available reasonable data evaluation pipeline (de Boni et al. 2011; Lutsik et al. 2011).

10.6.1 Bisulfite Treatment of Honeybee Genomic DNA

Here we describe two approaches for bisulfite treatment: (1) bisulfite treatment of genomic DNA packed in agarose beads and (2) bisulfite treatment of genomic DNA in solution. The first method is suitable for the treatment of small amounts of total genomic DNAs, while the second method is highly efficient for converting cytosines, but a loss of genomic DNA cannot be avoided. If enough genomic DNA is available for repeating experiments, the second method is preferable.

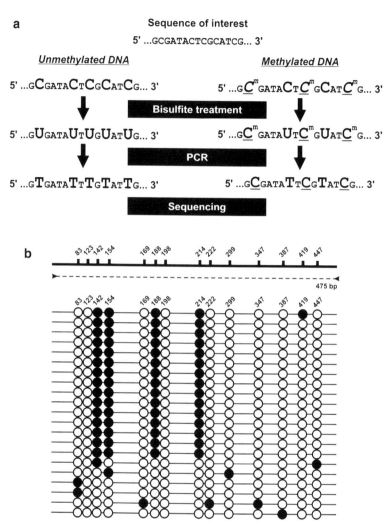

Fig. 10.4 (a) Principle of bisulfite PCR. (b) Diagrams showing the methylation status of a promoter region. *Filled boxes* on the *solid bars* indicate the candidate cytosine residues for methylation, and the number above each box indicates the nucleotide number counted from the first nucleotide of the PCR amplicon. *Arrowheads* and the *dotted line* represent primers and PCR amplicon, respectively. Results using a primer set within the element showed a mixture of methylated (*filled circles*) and unmethylated (*open circles*) cytosines. Cytosines 142, 154, 188, and 214 were methylated in about 70 % of sequences

1. *Bisulfite treatment of genomic DNA packed in agarose beads*: Five hundred nanogram of genomic DNA purified from honeybee brains was treated overnight with a restriction enzyme that does not cut within the target region to be amplified by PCR. Sodium bisulfite powder (3.8 g; mixture of $NaHSO_3$ and $Na_2S_2O_5$; Sigma) was dissolved in 5 mL distilled water and 1.5 mL 2 M NaOH, and 110 mg hydroquinone (Sigma) was dissolved in 1 mL distilled water and

mixed with sodium bisulfite solution. Next, 750 μL of the sodium bisulfite/hydroquinone solution was transferred to each tube and overlaid with 750 μL of heavy mineral oil. These aliquots were chilled on crushed ice for at least 30 min. Restricted DNA samples were denatured by boiling for 10 min and incubated with 0.33 M NaOH for 15 min at 50 °C. Two volumes of melted (50–55 °C) 2 % low-melting-point agarose gel (NuSieve® GTG® Agarose; Cambrex Bio Science Rockland Inc.) was mixed with DNA sample, and 25 μL of the sample solution was immediately transferred to the mineral oil layer of the chilled sodium bisulfite solution to form agarose beads by chilling in the ice-cold mineral oil layer. The tubes were left on ice for 30 min, allowing the agarose beads to sink into the bisulfite solution. Then, the samples were incubated at 50 °C for 3.5 h for bisulfite conversion. After briefly chilling on ice, all solutions were removed from the tubes, and agarose beads were washed subsequently with 1x TE buffer (2×15 min), 0.3 M NaOH (2×15 min), 1x TE buffer (10 min), and distilled water (10 min). Finally, the distilled water is completely removed, and the samples were stored at 4 °C.

2. *Bisulfite treatment of genomic DNA in solution*: Four hundred nanogram of genomic DNA, purified from honeybee brains, was treated overnight with a restriction enzyme that does not cut within the target region to be amplified by PCR. Sodium bisulfite powder (3.8 g; mixture of $NaHSO_3$ and $Na_2S_2O_5$; Sigma) was dissolved in 2.5 mL distilled water and 750 μL 2 M NaOH. Restricted DNA samples are mixed well with 187 μL of the bisulfite solution and 73 μL of scavenger chemical [98.6 mg 6-hydroxy-2,5,7,8-tetramethylchroman-2-carboxylic acid (Sigma) in 2.5 mL dioxane (Sigma)] and incubated in a PCR cycler at 99 °C for 15 min, at 50 °C for 30 min, at 99 °C for 5 min, at 50 °C for 1.5 h, at 99 °C for 5 min, and at 50 °C for 1.5 h. After adding 150 μL distilled water to the DNA samples, bisulfite-treated genomic DNA is purified and desulfonated using MICROCON® Centrifugal Filter Devices (YM-30 membrane; Millipore) by subsequent application of 500 μL 1x TE (2×15 min centrifugation), 50 μL 0.3 M NaOH (15 min incubation followed by 15 min centrifugation), and 1x TE buffer (15 min centrifugation). After elution using 50 μL of pre-warmed (50 °C) 1x TE, bisulfite-converted DNA is ready to use and can be stored at 4 °C.

Recently, kits for bisulfite treatment, which are designed to simplify and streamline the difficult procedures above, can be purchased from several companies, such as TaKaRa, QIAGEN, Invitrogen, and Millipore.

10.6.2 Bisulfite PCR and Cloning

PCR primer design for bisulfite-converted DNA is challenging but crucial for efficient amplification. Oligo primers should be designed to be devoid of CpG positions. In order to amplify specifically from bisulfite-converted DNA, all cytosines in forward primers and all guanines in reverse primers for this PCR were replaced with thymines and adenines, respectively. The length of the amplified PCR products should not exceed 650 bp. If the bisulfite treatment was performed with agarose beads (Method 1), the beads were melted by heating at 70 °C just before mixing into the PCR solution. PCR was performed using AccuPrime™ *Taq* DNA Polymerase System (Invitrogen) or Ex Taq™ polymerase (TaKaRa) for 40 cycles of 95 °C for 30 s, 50 °C for 30 s, and 68 °C for 1 min. High-fidelity DNA polymerases are recommended for the reactions. Following agarose gel electrophoresis and purification of PCR products from gels with the QIAquick Gel Extraction Kit (QIAGEN), the PCR products were inserted into a TA vector, such as pGEM®-T Easy vector (Promega). Competent *Escherichia coli* cells were transformed with cloned vectors and plated onto Luria-Bertani (LB) agar plates containing ampicillin and X-Gal/IPTG, which allowed the easy screening of positive transformants by appearance of white colonies. Positive white colonies were picked from the LB agar plates and grown in liquid LB medium. Plasmids with successful insertions were purified with QIAprep Spin Miniprep Kit (QIAGEN) and sequenced using primers which are located on the plasmid right upstream of the insertion (e.g., a T7 or SP6 primer in case pGEM®-T Easy vector was used).

10.6.3 Results of Bisulfite Sequencing

Twenty-one colonies were picked from the LB agarose plate for sequencing. Figure 10.4b represents the typical illustration of results of bisulfite sequencing, named lollipop scheme, to show the sites of 5-methylcytosine residues. We found four highly methylated cytosines in the amplified 475 bp of the promoter region (Fig. 10.4b). The amplified region was partially methylated in more than half of the sequences at the cytosines no. 142, 154, 188, and 214.

10.7 Single Nucleotide Primer Extension Assay

Although highly accurate, the described technique is still laborious and expensive, so samples should be well chosen before bisulfite sequencing is performed. Single nucleotide primer extension (SNuPE) technique can be regarded as a fast and cost-effective prescreening solution for quantitative analyses of methylation levels. This method is able to assess DNA methylation at one or two specific single cytosine residues representing the methylation state of the whole amplicon. Experimental protocol can be divided into four parts: (1) generation of PCR products derived from bisulfite PCR as described above, (2) purification of PCR amplicon, (3) SNuPE reaction, and (4) separation of the SNuPE products by high-performance liquid chromatography (HPLC) and quantification of the peaks (El-Maarri 2004). The SNuPE reaction is similar to Sanger sequencing reactions using unlabeled dideoxynucleotides which specifically detect the methylation status of the respective CpG in bisulfite PCR products (ddCTP and ddTTP when SNuPE primers are placed on

10 Epigenetic Regulation of Gene Expression in the Nervous System   161

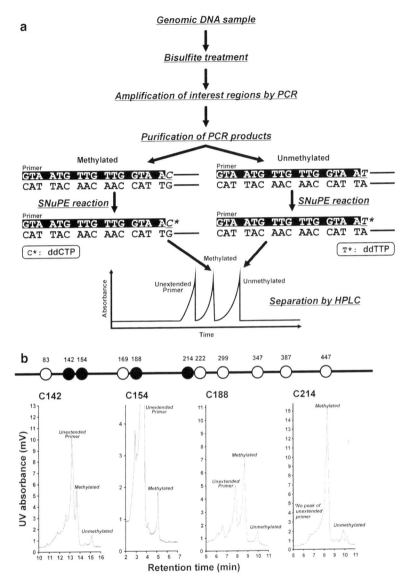

Fig. 10.5 (**a**) Principle of SNuPE analysis. (**b**) Measurement of methylation level of a promoter region by SNuPE analysis. UV spectra of SNuPE assays analyzing the methylation levels of cytosine 142 (C142), 154 (C154), 188 (C188), and 214(C214). Values in traces show the HPLC retention times of peaks

the top strand, ddGTP and ddATP when SNuPE primers are placed on the bottom strand) (Fig. 10.5a). Primers for an SNuPE reaction should be between 12 and 18 nucleotides long and are designed to match the site immediately adjacent to the cytosine residue of interest. Taken the T-rich top strand, ddCTP and ddTTP were added at the 3′-end of the primer depending on the methylation status of the target cytosine residue. The ddCTP and ddTTP were incorporated at the cytosine site which remains cytosines when methylated or show up as thymines when unmethylated, respectively, after bisulfite PCR. The ratio of the SNuPE product signals (C-extended primer and T-extended primer) is measured by HPLC.

10.7.1 SNuPE Reaction and HPLC

The procedure for SNuPE analysis has been modified from that previously described (El-Maarri 2004; Tierling et al. 2010). Bisulfite treatment of genomic DNA and subsequent PCR was performed using the same procedure described above for bisulfite sequencing. PCR products were purified with the QIAquick PCR Purification Kit (QIAGEN) or by Exonuclease I/Shrimp Alkaline Phosphatase (1 U ExoSAP) treatment. When larger PCR products are used, treatment with a restriction enzyme for 2 h, which digests the internal site of the PCR products, can be helpful to prevent forming of secondary structures and thereby inhibiting the SNuPE reaction. SNuPE primers were placed immediately adjacent to the target cytosine residues. The reaction was carried out in a total volume of 20 µl containing 100–130 ng purified PCR product, 1x buffer C (Solis BioDyne), 1.5 mM $MgCl_2$, 50 µM each ddNTP, 3.675 pmol of each primer, and 2.5 U TermiPol (Solis BioDyne) DNA polymerase, which shows high performance for incorporating ddNTPs. The thermal cycling was as follows: initial denaturing step of 96 °C for 2 min, followed by 50 cycles of 96 °C for 30 s, 50 °C for 30 s, and 60 °C for 60 s. Extension products were separated on an HPLC system (WAVE® DNA Fragment Analysis System; Transgenomic) at 50 °C and a flow rate of 0.9 mL/min using acetonitrile gradients, which were generated by increasingly mixing buffer B (0.1 M TEAA, 25 % acetonitrile) to buffer A [0.1 M triethylammonium acetate (TEAA)]: 25–37 % buffer B for 15 min for cytosine 142, 27–39% buffer B for 15 min for cytosine 154, and 24–32 % buffer B for 10 min for cytosine 188 and 214. After estimation of peak areas or heights, the methylation index (MI) was calculated as the ratio of the methylated signal divided by the sum of methylated and unmethylated signals.

10.7.2 Results of SNuPE Assay

We performed SNuPE analyses to exhaustively investigate the methylation level at the four sites of 5-methylcytosines identified by bisulfite sequencing (Fig. 10.5b). At cytosines 142 (C142), 188 (C188), and 214 (C214), large peaks indicating methylation were clearly observed together with small peaks representing unmethylated cytosines (Fig. 10.5b). These data clearly corresponded with the data obtained by bisulfite sequencing shown in Fig. 10.4b. We detected a peak of methylation at

10 Epigenetic Regulation of Gene Expression in the Nervous System

cytosine 154 but no peak of unmethylated cytosine (Fig. 10.5b). The amplitude of UV absorbance of the unextended primer for cytosine 154 was around 22, suggesting a weak primer extension efficiency (probably caused by minor amounts of PCR product).

10.8 Pharmacological Blocking of DNA Methylation

Many neurobiologists may have strong interests in the biological significance of control of DNA methylation status in brain function. In order to study this relationship, the pharmacological inhibition of DNA methylation has been utilized. In several studies, 5-azacytidine (Fig. 10.6a), 5-aza-2′-deoxycytidine (Fig. 10.6b), and zebularine (Fig. 10.6c) were directly injected into mammalian brains or honeybee hemolymph (Miller and Sweatt 2007; Miller et al. 2008; Han et al. 2010; Lockett et al. 2010). However, the molecular mechanisms of the action of azacytidines have

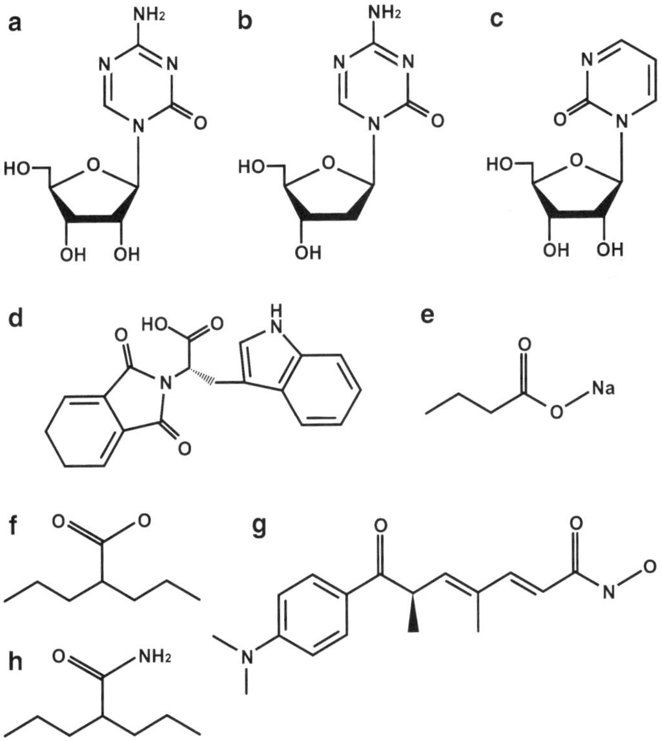

Fig. 10.6 Chemical structures of DNMT and HDAC inhibitors. (**a**) 5-azacytidine, (**b**) 5-aza-2′-deoxycytidine, (**c**) zebularine, (**d**) RG108, (**e**) sodium butyrate, (**f**) valproic acid, (**g**) trichostatin A, and (**h**) valpromide

to be considered. These chemicals are incorporated in newly polymerized genomic DNA molecules as false substrates in place of cytosines in the process of DNA replication during cell division and can act as DNMT inhibitors by trapping and inactivating DNMT molecules in the form of a covalent protein-DNA complex (Zhou et al. 2002; Lyko and Brown 2005). Azacytidines are effective DNA methylation inhibitors for frequently dividing cells, such as cancer cells (Hagemann et al. 2011) and stem cells (Balana et al. 2006). However, since mature neurons are not replaced and only very few neurons are incorporated into the existing neuronal circuitry by neurogenesis, azacytidines injected into brains may inhibit tRNA methylation (Schaefer et al. 2009). It has been suggested that methylation of tRNA regulates tRNA folding and stability (Alexandrov et al. 2006; Schaefer et al. 2010), and misfolded and unstable tRNA may affect the rate of protein synthesis.

Several non-nucleoside compounds can also be available as DNA methyltransferase inhibitors. One such inhibitor is (−)-epigallocatechin-3-gallate (EGCG), the main polyphenol compound found in green tea. EGCG inhibits the infectivity of influenza virus (Nakayama et al. 1993) and affects various biological processes in cancer (Kuzuhara et al. 2006, 2009; Siddiqui et al. 2011), including the blocking of DNA methyltransferase activity in cancer cells (Gu et al. 2009) and recombinant DNA methyltransferase protein activity (Rajavelu et al. 2011). Another DNA methyltransferase inhibitor, RG108 (Fig. 10.6d), inhibits the enzymatic activity of DNA methyltransferase by docking at the active pocket (Lyko and Brown 2005). This compound is effective without incorporation into genomic DNA and may be more suitable for neurobiological and behavioral experiments. In fact, intra-brain infusion of RG108 disrupts fear memory (Miller et al. 2010). The design and chemical synthesis of novel DNA methyltransferase inhibitors are useful in behavioral researches (Suzuki et al. 2010).

10.9 Methods for Analyzing Histone Modifications

Posttranslational modification of histones and chromatin remodeling are other essential epigenetic modifications that regulate gene expression. Histone modifications include acetylation, methylation, phosphorylation, ubiquitination, and ADP ribosylation (Levenson and Sweatt 2005; Schreiber et al. 2006). Acetylation is the best studied posttranslational modification of histone molecules, and acetylation of histones occurs on the amino group of the side chain of a lysine residue, resulting in effective neutralization of the positive charges of lysine. This modification dramatically alters the tertiary structure of chromatin to expose promoter regions, allowing greater access of transcriptional machinery, such as RNA polymerases and transcription factors, onto genomic DNA, resulting in enhancement of gene expression. Histone acetylation is catalyzed by histone acetyltransferases (HATs), which transfer acetyl groups from acetyl coenzyme A to the amino group of lysine side chains. HATs are also known to acetylate nonhistone proteins. This modification is reversible, and deacetylation is controlled by histone deacetylases (HDACs).

10.9.1 Pharmacological Blocking of HDACs

The enzymatic activity of HDACs is inhibited by applying their blocker molecules. HDAC inhibitors are categorized to five groups: (1) short-chain fatty acids, (2) hydroxamic acids, (3) cyclic tetrapeptides, (4) cyclic peptides, and (5) benzamides. In the field of neuroscience and neurobiology, *sodium butyrate* (a short-chain fatty acid; Fig. 10.6e; Bredy et al. 2007;), *valproic acid* (a derivative of a hydroxamic acid; Fig. 10.6f; Bredy et al. 2007; Rinaldi et al. 2007; Bredy and Barad 2008), and *trichostatin A* (a hydroxamic acid; Fig. 10.6g; Korzus et al. 2004; Levenson et al. 2004; Bredy et al. 2007; Chen et al. 2010) are well used to analyze interactions between histone acetylation and the consolidation of long-term memory. Sodium butyrate and trichostatin A are not only effective in mammals but also in invertebrates (crabs; Federman et al. 2009). Valproic acid inhibits GABA transaminase (Löscher 1993) and thus, by increasing the concentration of GABA in neurons (Czuczwar and Patsalos 2001), may interfere with neurotransmission of GABA. An analog of valproic acid, valpromide (Fig. 10.6h), also has anticonvulsant and mood-stabilizing effects but is not an HDAC inhibitor making it useful as a negative control for valproic acid (Bredy et al. 2007).

10.9.2 In Vitro Assays for Measurement of HAT Activity

As mentioned above, HATs add acetyl groups to lysine residues of histones and nonhistone proteins, and their enzymatic activity can be biochemically measured. For the broad quantification of HAT activity in a sample solution, the easiest way is to use HAT activity colorimetric assay kits (available from BioVision, Abcam, and other companies). By using the kit, HAT activity can be quantified by measuring the intensity of the yellowish color of the HAT substrate packaged in the kits. Another standard method for measuring HAT activity involves ^{3}H-marked acetyl coenzyme A. Samples are incubated with histones and ^{3}H-marked acetyl coenzyme A in a buffer containing 10 mM sodium butyrate (for blocking any contaminating HDACs in the sample) at 30 °C for 30 min. Then, sample solutions are dropped onto strong cation exchange papers (Grade P81, Whatman). The papers are washed with 0.2 M sodium carbonate solution buffer, and the intensity of radioactivity was measured with a scintillation counter.

Using ^{14}C-marked acetyl coenzyme A, target protein molecules of acetylation can be identified. After incubation under the same conditions described above, but substituting ^{3}H-marked acetyl coenzyme A with ^{14}C-marked acetyl coenzyme A, sample solutions were separated by standard SDS polyacrylamide gel electrophoresis. Gels were stained with Coomassie Brilliant Blue (CBB) and dried on filter paper, and the radioactivity was transferred onto imaging plates. By comparing the band patterns of CBB-stained gels and the radioactive images, acetylated proteins could be identified easily.

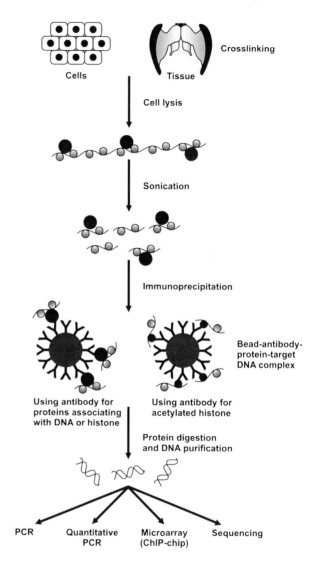

Fig. 10.7 Overview of chromatin immunoprecipitation methodology

10.9.3 Chromatin Immunoprecipitation

Chromatin immunoprecipitation (ChIP) is a powerful procedure for analyzing epigenetic modifications and to identify genomic DNA sequences bound to specific regulatory proteins, such as HAT proteins (Fig. 10.7). First, living cells or tissues are fixed with formaldehyde forming a strong cross-link between histones and genomic DNA. After shearing and solubilizing by sonication, pulldown is performed using an antibody for acetylated histones or proteins associated with genomic DNA and agarose/sepharose beads. Coprecipitated genomic DNA is resolved from histones by incubation in a high salt buffer and purified by phenol/chloroform treatment and ethanol precipitation.

Depending on the question addressed, the precipitated genomic DNA can be further analyzed by quantitative PCR, real-time PCR, next-generation sequencing, microarray, or other techniques. In a related microarray-based method, ChIP-on-chip, the precipitated DNA is labeled and hybridized to a variety of high-resolution microarrays.

10.10 Conclusion

All methods mentioned in this chapter have been modified from classical molecular biological and biochemical experiments and can be applied to any species of any insects. Honeybees, jewel wasps (*Nasonia vitripennis*), *Drosophila*, ants (*Pogonomyrmex barbatus*, *Linepithema humile*, and *Solenopsis invicta*), mosquitos (*Aedes aegypti* and *Anopheles gambiae*), pea aphids (*Acyrthosiphon pisum*), and red flour beetles (*Tribolium castaneum*) are insect species whose entire genome has been sequenced. Of these insect species, honeybees and ants have sophisticated and complex social behaviors. However, the epigenetic mechanisms involved in these behaviors remain unclear. The contribution of DNA methylation in the regulation of gene expression in insect taxa has yet to be elucidated, and detailed investigations combining standard and state-of-the-art methods, such as "next-generation" sequencing, are needed.

Recently, a species of sea slug *Aplysia* was shown to be useful for analyzing DNA methylation status at the single cell level (Moroz 2011; Moroz et al. 2011). The large diameter of molluscan neurons allows the copy number of specific mRNAs to be specifically measured by real-time PCR and pyrosequencing (Sadamoto et al. 2004; Hatakeyama et al. 2006; Moroz and Kohn 2010). Unfortunately, honeybees and other insects do not have such large neurons in their central nervous systems. Although molluscs do not display social behaviors like insects, the analysis of DNA methylation status in large single cells may spur the development of methods of investigating DNA methylation with ultrasmall amounts of genomic DNA.

References

Alexandrov A, Chernyakov I, Gu W, Hiley SL, Hughes TR, Grayhack EJ, Phizicky EM (2006) Rapid tRNA decay can result from lack of nonessential modifications. Mol Cell 21:87–96

Araujo FD, Knox JD, Ramchandani S, Pelletier R, Bigey P, Price G, Szyf M, Zannis-Hadjopoulos M (1999) Identification of initiation sites for DNA replication in the human dnmt1 (DNA-methyltransferase) locus. J Biol Chem 274:9335–9341

Balana B, Nicoletti C, Zahanich I, Graf EM, Christ T, Boxberger S, Ravens U (2006) 5-Azacytidine induces changes in electrophysiological properties of human mesenchymal stem cells. Cell Res 16:949–960

Ballester E, Pile LA, Wassarman DA, Wolffe AP, Wade P (2001) A *Drosophila* MBD family member is a transcriptional corepressor associated with specific genes. Eur J Biochem 268:5397–5406

Bock C, Reither S, Mikeska T, Paulsen M, Walter J, Lengauer T (2005) BiQ Analyzer: visualization and quality control for DNA methylation data from bisulfite sequencing. Bioinformatics 21:4067–4068

Bredy TW, Wu H, Crego C, Zellhoefer J, Sun YE, Barad M (2007) Histone modifications around individual BDNF gene promoters in prefrontal cortex are associated with extinction of conditioned fear. Learn Mem 14:268–276

Bredy TW, Barad M (2008) The histone deacetylase inhibitor valproic acid enhances acquisition, extinction, and reconsolidation of conditioned fear. Learn Mem 15:39–45

Chedin F, Lieber MR, Hsieh CL (2002) The DNA methyltransferase-like protein DNMT3L stimulates de novo methylation by Dnmt3a. Proc Natl Acad Sci USA 99:16916–16921

Chen G, Zou X, Watanabe H, van Deursen JM, Shen J (2010) CREB binding protein is required for both short-term and long-term memory formation. J Neurosci 30:13066–13077

Czuczwar SJ, Patsalos PN (2001) The new generation of GABA enhancers. Potential in the treatment of epilepsy. CNS Drugs 15:339–350

de Boni L, Tierling S, Roeber S, Walter J, Giese A, Kretzschmar HA (2011) Next-Generation sequencing reveals regional differences of the α-synuclein methylation state independent of Lewy body disease. Neuromolecular Med 13:310–320

Elango N, Hunt BG, Goodisman MA, Yi SV (2009) DNA methylation is widespread and associated with differential gene expression in castes of the honeybee, *Apis mellifera*. Proc Natl Acad Sci USA 106:11206–11211

El-Maarri O (2004) SIRPH analysis: SNuPE with IP-RP-HPLC for quantitative measurements of DNA methylation at specific CpG sites. Methods Mol Biol 287:195–205

Federman N, Fustiñana MS, Romano A (2009) Histone acetylation is recruited in consolidation as a molecular feature of stronger memories. Learn Mem 16:600–606

Field LM, Lyko F, Mandrioli M, Prantera G (2004) DNA methylation in insects. Insect Mol Biol 13:109–115

Foret S, Kucharski R, Pittelkow Y, Lockett GA, Maleszka R (2009) Epigenetic regulation of the honey bee transcriptome: unravelling the nature of methylated genes. BMC Genomics 10:472

Frommer M, McDonald LE, Millar DS, Collis CM, Watt F, Grigg GW, Molloy PL, Paul CL (1992) A genomic sequencing protocol that yields a positive display of 5-methylcytosine residues in individual DNA strands. Proc Natl Acad Sci USA 89:1827–1831

Gabor Miklos GL, Maleszka R (2011) Epigenomic communication systems in humans and honey bees: from molecules to behavior. Horm Behav 59:399–406

Glastad KM, Hunt BG, Yi SV, Goodisman MA (2011) DNA methylation in insects: on the brink of the epigenomic era. Insect Mol Biol 20:553–565

Goll MG, Kirpekar F, Maggert KA, Yoder JA, Hsieh CL, Zhang X, Golic KG, Jacobsen SE, Bestor TH (2006) Methylation of tRNAAsp by the DNA methyltransferase homolog Dnmt2. Science 311:395–398

Gu B, Ding Q, Xia G, Fang Z (2009) EGCG inhibits growth and induces apoptosis in renal cell carcinoma through TFPI-2 overexpression. Oncol Rep 21:635–640

Hagemann S, Heil O, Lyko F, Brueckner B (2011) Azacytidine and decitabine induce gene-specific and non-random DNA demethylation in human cancer cell lines. PLoS One 6:e17388

Han J, Li Y, Wang D, Wei C, Yang X, Sui N (2010) Effect of 5-aza-2-deoxycytidine microinjecting into hippocampus and prelimbic cortex on acquisition and retrieval of cocaine-induced place preference in C57BL/6 mice. Eur J Pharmacol 642:93–98

Hatakeyama D, Sadamoto H, Watanabe T, Wagatsuma A, Kobayashi S, Fujito Y, Yamashita M, Sakakibara M, Kemenes G, Ito E (2006) Requirement of new protein synthesis of a transcription factor for memory consolidation: paradoxical changes in mRNA and protein levels of C/EBP. J Mol Biol 356:569–577

Hoffman AR, Hu JF (2006) Directing DNA methylation to inhibit gene expression. Cell Mol Neurobiol 26:425–438

Hung MS, Karthikeyan N, Huang B, Koo HC, Kiger J, Shen CKJ (1999) *Drosophila* proteins related to vertebrate DNA (5-cytosine) methyltransferases. Proc Natl Acad Sci USA 96:11940–11945

Ikeda T, Furukawa S, Nakamura J, Sasaki M, Sasaki T (2011) CpG methylation in the hexamerin 110 gene in the European honeybee *Apis mellifera*. J Insect Sci 11:74

Jurkowski TP, Meusburger M, Phalke S, Helm M, Nellen W, Reuter G, Jeltsch A (2008) Human DNMT2 methylates tRNA(Asp) molecules using a DNA methyltransferase-like catalytic mechanism. RNA 14:1663–1670

Kamakura M (2011) Royalactin induces queen differentiation in honeybees. Nature 473:478–483

Kim KC, Friso S, Choi SW (2009) DNA methylation, an epigenetic mechanism connecting folate to healthy embryonic development and aging. J Nutr Biochem 20:917–926

Korzus E, Rosenfeld MG, Mayford M (2004) CBP histone acetyltransferase activity is a critical component of memory consolidation. Neuron 42:961–972

Krauss V, Eisenhardt C, Unger T (2009) The genome of the stick insect *Medauroidea extradentata* is strongly methylated within genes and repetitive DNA. PLoS One 4:e7223

Kucharski R, Maleszka J, Foret S, Maleszka R (2008) Nutritional control of reproductive status in honeybees via DNA methylation. Science 319:1827–1830

Kunert N, Marhold J, Stanke J, Stach D, Lyko F (2003) A Dnmt2-like protein mediates DNA methylation in *Drosophila*. Development 130:5083–5090

Kuzuhara T, Sei Y, Yamaguchi K, Suganuma M, Fujiki H (2006) DNA and RNA as new binding targets of green tea catechins. J Biol Chem 281:17446–17456

Kuzuhara T, Iwai Y, Takahashi H, Hatakeyama D, Echigo N (2009) Green tea catechins inhibit the endonuclease activity of influenza A virus RNA polymerase. PLoS Curr 1:RRN1052

Levenson JM, O'Riordan KJ, Brown KD, Trinh MA, Molfese DL, Sweatt JD (2004) Regulation of histone acetylation during memory formation in the hippocampus. J Biol Chem 279:40545–40559

Levenson JM, Sweatt JD (2005) Epigenetic mechanisms in memory formation. Nat Rev Neurosci 6:108–118

Locatelli F, Bundrock G, Müller U (2005) Focal and temporal release of glutamate in the mushroom bodies improves olfactory memory in *Apis mellifera*. J Neurosci 25:11614–11618

Lockett GA, Helliwell P, Maleszka R (2010) Involvement of DNA methylation in memory processing in the honey bee. Neuroreport 21:812–816

Löscher W (1993) In vivo administration of valproate reduces the nerve terminal (synaptosomal) activity of GABA aminotransferase in discrete brain areas of rats. Neurosci Lett 160:177–180

Lutsik P, Feuerbach L, Arand J, Lengauer T, Walter J, Bock C (2011) BiQ Analyzer HT: locus-specific analysis of DNA methylation by high-throughput bisulfite sequencing. Nucleic Acids Res 39:W551–W556

Lyko F, Foret S, Kucharski R, Wolf S, Falckenhayn C, Maleszka R (2010) The honey bee epigenomes: differential methylation of brain DNA in queens and workers. PLoS Biol 8:e1000506

Lyko F, Brown R (2005) DNA methyltransferase inhibitors and the development of epigenetic cancer therapies. J Natl Cancer Inst 97:1498–1506

Lyko F, Maleszka R (2011) Insects as innovative models for functional studies of DNA methylation. Trends Genet 4:127–131

Lyko F, Ramsahoye BH, Jaenisch R (2000) DNA methylation in *Drosophila melanogaster*. Nature 408:538–540

Marhold K, Kramer E, Kremmer F (2004) Luko, the *Drosophila* MBB2/3 protein mediates interactions between the MI-2 chromatin complex and CpT/A-methylated DNA. Development 131:6033–6039

Miller CA, Campbell SL, Sweatt JD (2008) DNA methylation and histone acetylation work in concert to regulate memory formation and synaptic plasticity. Neurobiol Learn Mem 89:599–603

Miller CA, Gavin CF, White JA, Parrish RR, Honasoge A, Yancey CR, Rivera IM, Rubio MD, Rumbaugh G, Sweatt JD (2010) Cortical DNA methylation maintains remote memory. Nat Neurosci 13:664–666

Miller CA, Sweatt JD (2007) Covalent modification of DNA regulates memory formation. Neuron 53:857–869

Moroz LL, Kohn AB (2010) Do different neurons age differently? Direct genome-wide analysis of aging in single identified cholinergic neurons. Front Aging Neurosci 2:1–18

Moroz LL (2011) Genomic deciphering of memory mechanisms and multiple origins of neural circuits. In: Abstracts of 12th symposium on invertebrate neurobiology, international society for invertebrate neurobiology, 52

Moroz LL, Citarella MR, Kohn AB (2011) Genomic portrait of a neuron: identification and quantification. Abstracts of 12th symposium on invertebrate neurobiology, international society for invertebrate neurobiology, 53

Müller U, Hildebrandt H (2002) Nitric oxide/cGMP-mediated protein kinase A activation in the antennal lobes plays an important role in appetitive reflex habituation in the honeybee. J Neurosci 22:8739–8747

Nakayama M, Suzuki K, Toda M, Okubo S, Hara Y, Shimamura T (1993) Inhibition of the infectivity of influenza virus by tea polyphenols. Antiviral Res 21:289–299

Novak P, Jensen T, Oshiro MM, Wozniak RJ, Nouzova M, Watts GS, Klimecki WT, Kim C, Futscher BW (2006) Epigenetic inactivation of the HOXA gene cluster in breast cancer. Cancer Res 66:10664–10670

Okada R, Rybak J, Manz G, Menzel R (2007) Learning-related plasticity in PE1 and other mushroom body-extrinsic neurons in the honeybee brain. J Neurosci 27:11736–11747

Park JM, Kunieda T, Kubo T (2003) The activity of Mblk-1, a mushroom body-selective transcription factor from the honeybee, is modulated by the Ras/MAPK pathway. J Biol Chem 278:18689–18694

Quint A, Cedar H (1981) In vitro methylation of DNA with Hpa II methylase. Nucleic Acids Res 9:633–646

Rae PM, Steele RE (1979) Absence of cytosine methylation at C-C-G-G and G-C-G-C sites in the rDNA coding regions and intervening sequences of *Drosophila* and the rDNA of other insects. Nucleic Acids Res 6:2987–2995

Rajavelu A, Tulyasheva Z, Jaiswal R, Jeltsch A, Kuhnert N (2011) The inhibition of the mammalian DNA methyltransferase 3a (Dnmt3a) by dietary black tea and coffee polyphenols. BMC Biochem 12:16

Rinaldi T, Kulangara K, Antoniello K, Markram H (2007) Elevated NMDA receptor levels and enhanced postsynaptic long-term potentiation induced by prenatal exposure to valproic acid. Proc Natl Acad Sci USA 104:13501–13506

Robinson KL, Tohidi-Esfahani D, Lo N, Simpson SJ, Sword GA (2011) Evidence for widespread genomic methylation in the migratory locust, *Locusta migratoria* (Orthoptera: Acrididae). PLoS One 6:e28167

Roder K, Hung MS, Lee TL, Lin TY, Xiao H, Isobe K, Juang JL, Shen CKJ (2000) Transcriptional repression by *Drosophila* methyl-CpG-binding proteins. Mol Cell Biol 20:7401–7409

Rodic N, Oka M, Hamazaki T, Murawski MR, Jorgensen M, Maatouk DM, Resnick JL, Li E, Terada N (2005) DNA methylation is required for silencing of ant4, an adenine nucleotide translocase selectively expressed in mouse embryonic stem cells and germ cells. Stem Cells 23:1314–1323

Sadamoto H, Sato H, Kobayashi S, Murakami J, Aonuma H, Ando H, Fujito Y, Hamano K, Awaji M, Lukowiak K, Urano A, Ito E (2004) CREB in the pond snail *Lymnaea stagnalis*: cloning, gene expression, and function in identifiable neurons of the central nervous system. J Neurobiol 58:455–466

Salzberg A, Fisher O, Simon-Tov R, Ankri S (2004) Identification of methylated sequences in genomic DNA of adult *Drosophila melanogaster*. Biochem Biophys Res Commun 322:465–469

Sanger F, Nicklen S, Coulson AR (1977) DNA sequencing with chain-terminating inhibitors. Proc Natl Acad Sci USA 74:5463–5467

Schaefer M, Hagemann S, Hanna K, Lyko F (2009) Azacytidine inhibits RNA methylation at DNMT2 target sites in human cancer cell lines. Cancer Res 69:8127–8132

Schaefer M, Lyko F (2007) DNA methylation with a sting: an active DNA methylation system in the honeybee. Bioessays 29:208–211

Schaefer M, Pollex T, Hanna K, Tuorto F, Meusburger M, Helm M, Lyko F (2010) RNA methylation by Dnmt2 protects transfer RNAs against stress-induced cleavage. Genes Dev 24:1590–1595

Schreiber V, Dantzer F, Ame JC, de Murcia G (2006) Poly(ADP-ribose): novel functions for an old molecule. Nat Rev Mol Cell Biol 7:517–528

Shi YY, Huang ZY, Zeng ZJ, Wang ZL, Wu XB, Yan WY (2011) Diet and cell size both affect queen-worker differentiation through DNA methylation in honey bees (*Apis mellifera*, Apidae). PLoS One 6:e18808

Siddiqui IA, Asim M, Hafeez BB, Adhami VM, Tarapore RS, Mukhtar H (2011) Green tea polyphenol EGCG blunts androgen receptor function in prostate cancer. FASEB J 25:1198–1207

Simmen MW, Leitgeb S, Charlton J, Jones SJM, Harris BR, Clark VH, Bird A (1999) Nonmethylated transposable elements and methylated genes in a chordate genome. Science 283:1164–1167

Suzuki MM, Kerr AR, De Sousa D, Bird A (2007) CpG methylation is targeted to transcriptional units in an invertebrate genome. Genome Res 17:625–631

Suzuki T, Tanaka R, Hamada S, Nakagawa H, Miyata N (2010) Design, synthesis, inhibitory activity, and binding mode study of novel DNA methyltransferase 1 inhibitors. Bioorg Med Chem Lett 20:1124–1127

Takeshima H, Suetake I, Shimahara H, Ura K, Tate S, Tajima S (2006) Distinct DNA methylation activity of Dnmt3a and Dnmt3b towards naked and nucleosomal DNA. J Biochem 139:503–515

Tang LY, Reddy MN, Rasheva V, Lee TL, Lin MJ, Hung MS, Shen CKJ (2003) The eukaryotic DNMT2 genes encode a new class of cytosine-5 DNA methyltransferases. J Biol Chem 278:33613–33616

Tierling S, Souren NY, Gries J, Loporto C, Groth M, Lutsik P, Neitzel H, Utz-Billing I, Gillessen-Kaesbach G, Kentenich H, Griesinger G, Sperling K, Schwinger E, Walter J (2010) Assisted reproductive technologies do not enhance the variability of DNA methylation imprints in human. J Med Genet 47:371–376

The Honeybee Genome Sequencing Consortium (2006) Insights into social insects from the genome of the honeybee *Apis mellifera*. Nature 443:931–949

Tweedie S, Charlton J, Clark V, Bird A (1997) Methylation of genomes and genes at the invertebrate–vertebrate boundary. Mol Cell Biol 17:1469–1475

Tweedie S, Ng HH, Barlow AL, Turner BM, Hendrich B, Bird A (1999) Vestiges of a DNA methylation system in *Drosophila melanogaster*? Nat Genet 23:389–390

Wang Y, Jorda M, Jones PL, Maleszka R, Ling X, Robertson HM, Mizzen CA, Peinado MA, Robinson G (2006) Functional CpG methylation system in a social insect. Science 314:645–647

Watanabe T, Kikuchi M, Hatakeyama D, Shiga H, Yamamoto T, Aonuma H, Takahata M, Suzuki N, Ito E (2006) Gaseous neuromodulator-related genes expressed in the brain of honeybee *Apis mellifera*. Dev Neurobiol 67:456–473

Zhou L, Cheng X, Connolly BA, Dickman MJ, Hurd PJ, Hornby DP (2002) Zebularine: a novel DNA methylation inhibitor that forms a covalent complex with DNA methyltransferases. J Mol Biol 321:591–599

Index

A

Absolute conditioning, 17
Acetoxymethyl (AM) ester, 71
Acquisition, 15
Activity-dependent genes, 124
Activity-dependent labeling, 90
Antennal lobes, 26
Anxiety, 112
Appetitive learning, 27
Arc, 124
Associative conditioning, 19
Associative learning, 14
Aversive learning, 27
5-azacytidine, 153
5-aza-2'-deoxycytidine, 153

B

Bayes' Rule, 55, 56
Binary expression system, 97
Bisulfite PCR, 146–150
Brain, 35
Brain-machine interfaces (BMIs), 36
Brain sectioning, 126

C

Ca indicators, 89
Calmodulin, 7
Cameleon, 2, 103
Ca^{2+} transient, 67
c-fos, 124
Chromatin immunoprecipitation, 156–157
Chromophore, 99, 100
Classical conditioning, 14
Complementary metal-oxide-semiconductor (CMOS), 73
Conditional probability, 55
Conditioned stimulus (CS), 14
Confocal laser-scanning microscope (CLSM), 75
Connectivity, 38
CpG, 143–145, 149, 150
Cryostat, 126
Cytosine methylation, 144

D

Delivery, 116–117
Dextran-conjugated Ca^{2+} indicator, 71
Differential conditioning, 23
DNA methyltransferases (DNMTs), 142–144, 154
Drosophila melanogaster, 97

E

Earthworm, 84
Effective connectivity, 42
EGCG. *See* (−)-epigallocatechin-3-gallate (EGCG)
Egr1, 124
Electron-multiplying CCD (EM-CCD), 73
(−)-epigallocatechin-3-gallate (EGCG), 154
Epigenetic, 142–144, 146, 154, 156, 157
Excitatory, 113

F

FM1-43, 90
FMRF amide, 86
Förster resonance-energy transfer (FRET), 4, 99, 100
Functional connectivity, 32

G
GAL4, 97, 98, 100, 103
Gaussian random variable, 57
GCaMP, 2, 99, 103
GECIs. *See* Genetically encoded Ca^{2+} indicators (GECIs)
Generative model, 55, 57
Genetically encoded Ca^{2+} indicators (GECIs), 99, 100, 102, 103
Granger causality (GC), 43

H
Histone acetyltransferases (HATs), 154–156
Histone deacetylases (HDACs), 154, 155

I
Imaging, 100–103
Immediate early genes (IEGs), 124
Immunohistochemistry, 86
Inhibit(ing/ory), 113, 115
In situ hybridization, 124
Interaural time difference (ITD), 56
Intracellular signaling, 115–116

K
K_d dissociation constant, 67

L
Learning, 14
Likelihood function, 55–57
Lipophilic fluorescent dyes, 85
Locomotory behavior, 88
Long-term memory (LTM), 26

M
MAP estimate, 55, 60
Marginal distribution, 55
Maximum likelihood, 55
MBD. *See* Methyl-CpG-binding domain proteins (MBDs)
Medium-term memory (MTM), 25
Memory, 14, 112
Memory retrieval, 15
Methyl-CpG-binding domain proteins (MBDs), 143, 144
Molecular, 113–117

Motor, 111–112
Mushroom bodies, 26

N
Neural correlations, 62
Neuroengineering, 36
Neuroscience, 36
Nipkow-type spinning-disk microscope niokow-disk microscope, 75
Nitric mono-oxide (NO), 85
NO-specific fluorescent dye, 87

O
Olfactory learning, 14
Optical, 117–118
Optogenetics, 108–109

P
Pavlovian conditioning, 14
PER. *See* Proboscis extension reflex (PER)
Poisson neurons, 59, 60
Population vector, 58, 61
Posterior distribution, 55, 58
Prior distribution, 55, 57, 58
Probabilistic population code, 61
Proboscis extension reflex (PER), 14, 15, 17, 19, 21–23, 25–28

R
Ratiometric, 99, 100
Ratiometric dye, 68
Respiration, 110
Retina, 109–110
RG108, 154

S
Short-term memory (STM), 25
Singlemetric dye, 67
Single nucleotide primer extension (SNuPE), 150–153
Sleep, 110–111
SNuPE. *See* Single nucleotide primer extension (SNuPE)
Sodium butyrate, 155
Southern-Western blotting, 145
S^{35}-RNA probe, 129–130
Stimulating, 113–114

T
Thermotaxis, 4
Time-frequency representations, 38
Trichostatin A, 155
Tuning curve, 60
Two-photon microscopy (TPM), 75

U
Unconditioned stimulus (US), 14

V
Valproic acid, 155
Valpromide, 155
Visual learning, 14
Voltage-dependent fluorescent
    dyes, 90

Z
Zebularine, 153

Printed by Publishers' Graphics LLC
LMO130726.15.14.44